STUDIES IN PHILOSOPHY

Edited by
Robert Nozick
Pellegrino University Professor
at Harvard University

A GARLAND SERIES

ESSAYS ON SYMMETRY

Jenann Ismael

Garland Publishing, Inc.
New York & London / 2001

Published in 2001 by
Garland Publishing, Inc.
29 West 35th Street
New York, NY 10001

Garland is an imprint of the Taylor & Francis Group

Copyright © 2001 by Jenann Ismael

All rights reserved. No part of this book may be reprinted or reproduced or utilized in any form or by any electronic, mechanical, or other means, now known or hereafter invented, including photocopying and recording, or in any information storage or retrieval system, without permission in writing from the publisher.

10 9 8 7 6 5 4 3 2 1

Library of Congress Cataloging-in-Publication Data
Ismael, Jenann, 1968–
  Essays on symmetry / Jenann Ismael.
    p. cm. — (Studies in philosophy)
  Originally presented as the author's thesis (doctoral)—Princeton University.
  Includes bibliographical references and index.
  ISBN 0-8153-3603-9 (alk. paper)
  1. Symmetry. I. Title.
Q172.5.S95 I85 1999
501—dc21   99-045884

Printed on acid-free, 250-year-life paper.
Manufactured in the United States of America

To my parents,
with admiration and love.

And to my sister,
who gave them two grandchildren,
while I did this.

Contents

Abstract	ix
Preface	xi
Acknowledgments	xiii
Introduction	1
1. Framework	3
2. Space-Time Theories	5
3. Essays 1 and 2	9
4. Essays 3 and 4	10
5. Essay 5	16
Essay 1: Symmetry: an Elementary Introduction	19
1. Introduction	19
2. Geometric Symmetries	19
3. Finite Groups in One and Two Dimensions	23
4. Recap and Generalization	27
Essay 2: Curie's Principle	31
1. Introduction	31
2. The Principle	32
3. The Proof	34
4. Interpretation of the Principle	37
5. How One Could Doubt the Principle	39
6. The Principle in Indeterministic Contexts	42
7. Spontaneous Symmetry Breaking	45
8. The Principle and Hidden Variables	47

9. Exegetical Support — 49
10. Is the Principle Trivial? — 52
11. Conclusion — 53
Appendix: Geometric Transformations — 55

Essay 3: Reflection in Space — 59

1. The Interdefinability of Symmetry, Similarity, and Intrinsic Properties — 59
2. Enantiomorphs — 61
3. Kant, Remnant, Earman, and Nerlich — 64
4. A Better Argument — 72
5. Symmetry as a Constraint on Meaning — 77
6. Parity Violation — 78

Essay 4: Asymmetry — 85

1. Introduction — 85
2. Distinction between Laws and Initial Conditions — 86
3. Relation between the Asymmetries of the Laws and the Asymmetries of Initial Conditions — 87
4. Initial Conditions — 89
5. Reflection in Space — 90
6. Reflection in Time and Thermodynamics — 94

Essay 5: Science and Symmetry — 107

1. Introduction — 107
2. Structure of Theories — 108
3. Observable/Unobservable Distinction — 120
4. Content of Theories: Interpretation — 136
5. *Why* do we Discriminate Between Empirically Equivalent Theories? — 146
6. *How* do we Choose between Empirically Equivalent Theories? — 162
7. Properties — 183

Bibliography — 199

Author Index — 205

Subject Index — 207

Abstract

Structures of many different sorts arise in physics, e.g., the concrete structures of material bodies, the structure exemplified by the spatiotemporal configuration of a set of bodies, the structures of more abstract objects like states, state-spaces, laws, and so on. To each structure of any of these types there corresponds a set of transformations (appropriate to structures of the relevant type) which map it onto itself. These are its symmetries.

Increasingly ubiquitous in theoretical discussions in physics, the notion of symmetry is also at the root of some time-worn philosophical debates. The dissertation consists of a set of essays on overlapping topics drawn from both fields. The first is an elementary and informal introduction the mathematics of symmetry. The second is a discussion and defense of a famous principle of Pierre Curie which states that the symmetries of a cause are always symmetries of its effect. The third essay takes up the case of reflection in space in the context of a controversy stemming from Kant's attempt to 'prove' that space is a substance on the basis of the observation that certain material objects cannot be brought into coincidence with their images under reflection by any continuous rigid motion. The fourth essay considers the case of reflection in time against the background of a discussion of the general conditions under which phenomena which are asymmetric with respect to a given transformation suggest that the laws governing the phenomena are asymmetric with respect to the transformation in question. The theoretical strategy followed by statistical mechanics in accommodating the temporally asymmetric phenomena which fall under the Second Law of Thermodynamics to the temporally symmetric classical dynamical laws is used to illustrate the general points. The long fifth essay articulates a view about the nature of scientific theorizing, one that is suggested by the abstract mathematical perspective encouraged by attention to symmetries and is implicit in the preceding essays.

Preface

No one should be forced to read their dissertation until there is enough of a distinguished career between they and the childish author of the thing that they don't feel too acutely their own judgement of it, i.e., that they don't experience it as particularly *self-directed*. Either that, or they should be just finished, still protected from too clear a view of it by infatuation (with their ideas) and ignorance (of those of others). Casting a cold eye from this distance—just three years away and with not so much as a beard and some modest professional accolades to buffer the affect of sober assessment—is more painful than I would have imagined. I have resisted, nevertheless, the temptation to make more than cosmetic changes, and the thesis stands, for what it's worth, as an honest record of immature thought

The reasons I chose to write about symmetry are easy to state: its increasing ubiquity in theoretical discussion in physics, its presence not only in the ancient philosophical contexts I discuss, but in some of the most influential contemporary debates (Quinean arguments for ontological relativity and the indeterminacy of reference, Putnam's model-theoretic argument, Poincare's and Reichenbach's arguments for the conventionality of geometry: these, and too many others to list, are arguments from symmetry). All of these things contributed to a suspicion that there was much to be learned from watching how symmetries operate in thought about the physical world. There were aesthetic reasons too: the sheer beauty of the mathematical theory, and something in the abstractness of the perspective that appealed to me. These still stand, and I still can't think of a topic in philosophy that has received so little of the attention it deserves.

Acknowledgements

My philosophical debt Bas van Fraassen is evident on every page of this dissertation, to he has also had an influence of a different sort, which is harder to gauge and for which I am at least as grateful.

To David Albert I owe endless thanks for a wholly unexpected generosity and encouragement, and for the simple beauty of his work.

To Elijah Millgram and Gideon Rosen I am indebted for reading parts of this material, and to Michael Elowitz I am indebted for the ample evidence the thesis bears of the many hours spent away from it. He made me happy.

A very special and very personal thank you is due to Paul Benacerraf ... for guidance and (I hope) friendship.

My parents I cannot thank enough; I owe them, and love them, more than I can express.

Introduction

There's a kind of *gestalt* shift encouraged by science; to see the world as portrayed by the models of our physical theories is to see it in terms of properties quite far removed from the throng of colors, smells, sounds, and the like given immediately in experience. Attention to the former reveals surprising regularity folded into the apparent disarray of the visible world. In essay 5, I use the example of music to illustrate how simplicity of materials can be combined with complexity of construction to produce elaborate patterns out of elements drawn from a very small stock of types. A Beethoven sonata, a Handel concerto, Bach's *Well-Tempered Clavier*; all of these—and all compositions expressible in canonical musical notation which anyone has ever conceived or ever will — are comprised of notes and relations drawn from the same meager stock, they differ only in their number and arrangement. The world, as modeled by our favored theories, is just such a structure. The ostentatious variety it exhibits to the senses is nothing but the artful product of intricacy of construction; in truth, there are only a very few basic types of thing and a small number of types of relation they bear one another. Even of the mathematically describable constructions out of these elements, only a relatively small subset are physically realizable, because only a relatively small subset satisfy the physical laws.

There are any number of families of properties in terms of which we might represent the world; there is the pattern formed by the distribution of visual qualities, another formed by the distribution of smells, another formed by the distribution of grue-style gerrymanderings of visual qualities, another formed by things whose English names begin with the various letters of the English alphabet (granted some convention for resolving ambiguity in case a thing has more than one description) ... and so on indefinitely. What is special about the properties or quantities which we take as basic in our physical theories, i.e. what distinguishes 'having mass m', 'having energy E', and 'being positively charged' from 'being pink', 'being grue', and 'having an English name beginning with "r"'? I suggest that it is nothing intrinsic to mass, energy, or charge which makes them especially suited to the role of basic quantities in theorizing; they have no special ontological status, nor do they correspond to universals or 'carve Nature at her Joints'.[1] It is rather that the pattern formed by their actual distribution has features which make a great deal of it easy to guess from knowledge of a small part, and that we can obtain knowledge of such a part from experience. If this is right, then it is not the *properties* but their *instantiation patterns* in the actual world which we are most interested in. The instrumental aim of theorizing is the construction of models which accurately represent the world; it is served by fixing on the set of properties whose actual instantiation pattern can be most fully and accurately determined on the basis of the partial knowledge observation provides us. So, a large part of theorizing is a matter of canvassing the actual instantiation patterns of various sets of properties, and this is the source of the scientific interest of symmetry, for there is no more elegant and revealing way of studying such patterns than in terms of their symmetries.

Besides theoretical models, other patterns or structures (in particular, laws and states) play important roles in physics, and the symmetries of these are intimately connected with one another and with the symmetries of theoretical models.[2] One of the main things I hope to illustrate is how revealing it is to view these in terms of their symmetries, and how simply and saliently the important relations between them, and between those associated with different theories, appear when represented in terms of their symmetries. We will need, in all of this, a general framework in terms of which we can present the content of an arbitrary theory,

---

1. This is overstated. Perhaps we can infer from their suitability as values for fundamental quantities to their having a special ontological status; the point here is that it is not any such status which determines their suitability for their role in science.
2. I use the terms 'pattern' and 'structure' interchangeably.

Introduction

and I'll sketch one only at the level of detail necessary to bring out the features of theories I will want to focus on.[3]

1. FRAMEWORK

A physical theory is associated with an ontology and a set of laws. The ontology specifies the basic individuals, the families of properties—or 'quantities'—whose values characterize them intrinsically and the external relations they bear to one another. In so doing, it delimits the set of worlds which are metaphysically or 'materialistically' possible: those obtained by recombination of properties and relations pertaining to actual individuals. Intuitively, these are just worlds made up of the same stuff as the actual world, differently arranged. Among these, worlds which satisfy the laws of a given theory are its **models**. It will be most convenient for us to conceive of a theory as an ordered pair <M,P> where **M** is a set of materialistically possible worlds and **P**, the subset of these which constitute models of the theory. Theories are distinguished either by their associated sets of materialistically possible worlds or by the subsets of these which are regarded as physically possible. Features which play an important role in some conceptions of theories, such as the language in which theories are formulated or the mathematical form of the equations which express their laws, are here regarded as incidental and relegated to the subsidiary role of picking out the sets of worlds which are the real locus of interest.

We can give simple definitions in these terms of the notions that will most occupy us: a **transformation**, h, is something like a set of instructions for reorganizing an arbitrary world, it tells us how to replace certain types of objects and events with others and move things around in space and time.[4] The **symmetries of a world M** are transformations which have no real effect on **M**. If h is a symmetry of **M**, we can spend the whole day reorganizing M in

---

3. Here and throughout, I will be unabashed in suggesting picturesque analogies to help think about one or another point. When rigor is required pictures cannot be relied on, but there is nothing wrong with exploiting them in the meantime. Indeed, much informal reasoning consists in manipulating such mental images, and part of what makes symmetry such a powerful conceptual tool is precisely that it lends itself so naturally to simple images.
4. It will become clear when we give a more precise definition of transformation that a transformation is a certain kind of function, and that it is only 'something like' a set of instructions because it may not be computable.

accordance with the h-instructions to no real effect; what we will have at the end of the day is simply a duplicate of **M**. The transformations which are not symmetries of **M**, by contrast, leave M changed; once we have finished reorganizing a world in accordance with the directions supplied by a transformation which is not a symmetry of M we have a world which is different from **M**.

The **symmetries of a theory** are something different from the symmetries of the individual worlds which constitute its models.[5] Whereas the symmetries of a world preserve all of the world's properties, the symmetries of a theory, **T**, in general preserve only a single property: that of satisfying, or failing to satisfy, **T**. If h is a symmetry of T, h maps the set of T-models onto itself; it may change a world in whatever way you like so long as it never changes a T-model into a non-model, or a non-T-model into a model.[6] The physical significance of the symmetries of a theory is easily conveyed with a very little theory, **T**, which gives the value of one quantity as a function of the value of another, and is expressed by the simple formula $A=f(B)$.[7] The only transformations which are not symmetries of T are (by definition) those that map some T-solution onto a non-T-solution or vice versa. This happens when and only when the transformation in question induces a change in the value of A in a world without making compensating changes in the value of B, so the only transformations which are not symmetries of **T** are those which permute the values of A without always counterbalancing the effect by inducing appropriate changes in the B-values. The transformations which are not symmetries of the B-determining laws of **T** (i.e. their 'asymmetries'), then, can be thought of as inducing dynamically relevant changes in the values

---

5. There is one exception: if a theory has no models or only one model, then its symmetry group and that of its models will coincide.
6. h must also map no two different models of T onto the same T-model. I will sometimes leave this clause off, though it always applies.
7. A and B need not be basic quantities; they may be functions of sets of such quantities. The equation of state in thermodynamics, to take a simple example, relates the product of the pressure and volume of an ideal gas to its temperature multiplied by a constant (which depends on state-independent properties of the gas in question), $PV=kT$. The expressions on both sides of the equation denote quantities, but neither are basic dynamical quantities. If we allow that any function of the basic quantities characterizing a system is itself a well-defined quantity for the system, all laws have this form.
If the theory in question is indeterministic, B may not be a quantity but a probability distribution over values of a quantity, or a set of possible values, with no probability distribution. More is said about quantities in essay 5.

Introduction

of quantities which are nomological determinants of B, or, less clumsily, as permutations of the values of quantities that are dynamically relevant to B. The symmetries of the B-determining T-laws are, by contrast, permutations of the values of quantities which are dynamically irrelevant to B.[8]

In general, the equations of a physical theory give the value of one quantity as a function of the value of another.[9] Since one can't change the value of B while preserving the functional dependence of B on A unless one makes compensating changes in the value of A, if a given transformation is a symmetry of the B-determining laws it must *either have no net effect on A or counterbalance the effect it has by making compensating changes in the value of B*. So, while the symmetries of a model map the model onto itself, the symmetries of a theory map all and only models of the theory onto models.

2. SPACE-TIME THEORIES

We can make this all a bit more mathematically precise if we concentrate on the class of theories that can be formulated as what Friedman and Glymour call 'space-time theories'. The basic entity of such theories is the collection of space-time points or 'places-at-a-time', and a materialistically possible world is completely described by specifying the intrinsic properties of the points and the network of external relations between them. The former are

---

8. Suppose A is the product of two more basic real-valued quantities, C and D, and h is the transformation which doubles the value of C while halving that of D. Since A is dynamically relevant to B, if we regard C and D as also dynamically relevant to B, we will have to recognize h as a permutation of the values of quantities which are dynamically relevant to B, but it is not an asymmetry of the B-determining laws.

  We have two choices. One is to qualify our earlier statement by saying that the asymmetries of the B-determining laws are dynamically relevant permutations of the values of dynamically relevant parameters. The other is to deny that C and D are individually dynamically relevant to B, although their product, A, is. In this case, we can still hold that the asymmetries of the B-determining laws are those permutations of quantities which are dynamically relevant to B, since h doesn't permute the value of A. This way of speaking, though less natural than allowing that the more basic constituents of a complex quantity which is dynamically relevant to B, are also dynamically relevant to encouraging us to think of quantities in the way I think is best, viz. to treat all quantities (i.e. arbitrary functions on phase space) as on an equal footing, without presupposing any absolute division between simple and complex quantities, and without any privileged basic set of quantities. See section vii of essay 5.

9. Laws of coexistence give the value of one quantity at a single time as a function of some other, deterministic dynamical laws give the state of a system at a time as a function of its state at others, indeterministic laws are a little different, and I'll leave them aside for the moment.

given by the values the various field quantities take at points, and the latter by the spatiotemporal structure they describe. It is assumed that the points have the minimal structure of a manifold so that additional spatiotemporal structures and matter fields can be specified in a simple manner by geometric objects defined thereon. The laws of these theories take the form of field equations relating source variables to field quantities and equations of motion which pick out the sets of events which represent physically possible trajectories of material systems.[10]

As Glymour remarks, this style of formulation is "natural in the sense that actual theories are sometimes stated that way, they are reasonably clear, and we know how to write down a great many theories in such terms." ("Epistemology of Geometry", in *Theory and Evidence*, p. 344).[11] Its strongest recommendation is its clarity; we can tell what a theory says, and in particular what it says about the spatiotemporal structure of the world, when it is formulated in such a manner. This distinguishes it from the more traditional style of formulation on which the laws of a theory are stated in terms of, and hold only relative to, a restricted class of coordinate systems. On the coordinate-dependent style of formulation, one is presented with a description of the models of a theory relative to one coordinate system, told how to translate each into a description of a model of the theory relative to any other of the coordinate systems in a restricted class (those obtained from the original by transformations in the so-called covariance group of the theory), and left to sort out for oneself which of the spatiotemporal structures belong to a given model and which are artifacts of the coordinate representations in terms of which they are given. An (imperfect but useful) analogy is provided by thinking of a coordinate system as a kind of lens through which worlds are viewed. On a coordinate-dependent formulation, one is told how the models of a theory look through one lens, and told which additional lenses are such that all and only worlds which appear as models under the original lens will also appear as models under them. It is not guaranteed that a given world appear under such a

---

10. This is just a skeleton; there may be, in addition, equations establishing boundary conditions under certain circumstances, symmetry principles (e.g., symmetric sources have symmetric fields), and the like.
11. The approach is most fully developed in Friedman's Foundations of Space-time Theories (Princeton University Press, Princeton, NJ).

lens as the *same* sort of world as it did under the original, but if it appeared (failed to appear) as a model under the original lens, it is certain to appear (fail to appear) as such under the new one. Whatever effect switching among a theory's associated lenses has, it doesn't turn its models into non-models or *vice versa*.[12] It is not hard to see that determining from this what the various individual models look like *independently of the lenses* is going to be a messy business, and historically it has made for no small amount of confusion about the spatiotemporal structure associated with the models of different theories, and the dynamical role played by that structure. To agree, once again, with Glymour's assessment, it is better by far to describe the content of a theory in the more straightforward manner of space-time theories, where we know how to do it:

> Although any theory may imply, given a coordinate system, various coordinate dependent equations, and may sometimes be more easily tested in such a form, it is essential that the theory be *stated* [i.e. in the coordinate independent style of a space-time theory]; otherwise we become enormously confused about what each theory claims, about the synonymy of theories, and so on. ("Epistemology of Geometry", in *Theory and Evidence*, p. 344).

For these reasons, I will adopt the framework of space-time theories, making the minimal structural assumptions about the models of a theory that allows us to pour them into that mold. Friedman has given space-time formulations of the important classical theories (Newtonian Dynamics, Special Relativity, and General Relativity) and the expectation in physics nowadays is that whatever the true theory turns out to be, it will be formulable in this way. So whatever generality is lost by restricting attention to space-time theories, it will not rule out any of the actual or currently envisioned physical theories which are our principal concern, and the precision and clarity that is gained thereby are worth it.

---

12. This is not to say that it doesn't, as I mentioned, have effects; it may, for instance, turn all red manifolds green and all green manifolds red, while turning blue manifolds into pink ones and pink ones into blue. So long as both or neither of these manifolds in each of these pairs are models, it will be fine. What it can't do, if green manifolds are models and pink ones not, is make manifolds that looked green under the original lens appear pink.

Adopting a space-time framework means making some assumptions about ontology, specifically, assuming that each materialistically possible world consists of a manifold of points (or, if you prefer, events; an event is the instantiation of the value of a quantity at a point). To say that the points form a manifold is to say that they have a topology and that they are *coordinatizable* by R^4 (i.e. given any point p, there is a notion of a neighborhood A of p and a one-one map F (a 'chart' around p) from A into R^4 that takes nearby points in A onto 'nearby' points in R^4 and *vice versa*). The assumption allows us to translate statements about geometrical entities in space-time into statements about real numbers (for instance, to describe arbitrary curves by real-valued functions), and it provides us with a natural notion of differentiability (if f is a real-valued function defined on neighborhood in space-time, it is differentiable just in case $f.F^{-1}$ is differentiable for every chart F). Granted this, the additional geometrical structure of, and the matter fields in, space-time can be described by geometric objects (tensor fields or affine connections) defined on the manifold, and we can represent possible worlds by n-tuples $<M, F_1, F_2 ... F_m>$ (where M is a four-dimensional manifold, $F_1, F_2 ... F_m$ are geometrical objects defined on M), and theories by classes of such n-tuples corresponding to worlds which satisfy its laws.[13] The types of matter field and the level of geometric structure attributed our world varies from theory to theory; the particular field configurations and certain of the geometric structures (those which are not 'absolute') vary from model to model of a given theory.

So, a theory consists of an ordered pair T=<**M,P**>, where **M** is a set of models representing worlds which are materialistically possible according to T, **P** is a subset of **M** representing worlds which are physically possible according to T, and each of the models has the structure of a manifold of events, i.e. each element in **M** and **P** is an n+1-tuple $<M, F_1...F_n>$ where M is a four-dimen-

---

[13]. Besides the intrinsic geometric structures introduced to account for the trajectories of free particles, the geometrical objects include those introduced to account for the motions of particles acted on by external forces: source variables representing the sources of the interaction and field variables representing the forces arising from the interactions. Field equations relate the source variables to the field variables and equations of motion pick out the trajectories of particles effected by the force in question.

Introduction 9

sional differentiable manifold and $F_1...F_n$ are geometric objects representing spatiotemporal structures and matter fields defined thereon.

We won't make too much use of this *technicalia*, but it is important because it provides us with a precise notion of transformation and gives us a grip on the constitution of the set of transformations which apply to the kinds of worlds we will be considering. The transformations in question are *manifold automorphisms*, one-one suitably bicontinuous and differentiable mappings of every neighborhood of **M** into **M**. A transformation on M maps any geometrical object Q defined on M onto another object hQ; the **symmetries of a geometric object Q** are transformations which map it onto itself; the **symmetries of a model M** itself are transformations which are symmetries of all of the geometric objects defined on M; and the **symmetries of a theory T** (of which M may or may not be a model) are automorphisms of the set of T-models, transformations which map all and only T-models onto T-models.

3. ESSAYS 1 AND 2

It takes a little while to become accustomed to thinking with this apparatus, and the first essay is devoted to familiarizing it by introducing the notion of symmetry in a more intuitive way with some simple geometric examples, generalizing it to non-geometric mathematical contexts, and finally paving the way for discussion of its applications in physics. Once one is adept with it, however, the apparatus is rather astonishingly powerful. Its power derives from its generality: various kinds of structures play central roles in theorizing (the structure of the phenomena, of the models of a theory, of a theory's laws, and so on), and we can formulate some surprisingly general and important principles concerning the relations between the symmetries of these structures.

The second essay provides quite a nice illustration of this in an often misinterpreted and unjustly maligned principle put forward by Pierre Curie in 1894. The principle is a simple mathematical truth that scarcely needs proof once it is properly understood. It says that if we are given two sets, with a one-many mapping of the one into the other, the transformations which are symmetries of every element of the former are also symmetries of every element of the latter.[14] In slogan form, and in the ill-advised causal terminology in terms of which Curie expressed it, *the symmetries of a cause are preserved in its effect*.[15] Special, and especially important, cases of the principle include those in which (i) the structures in the two sets are state descriptions (or partial state descriptions) and the mapping is specified by a physical law, (ii) the structures in the two sets are properties (e.g., a proposed reduction basis and the properties to be reduced, respectively) and the mapping is the philosophical analysis which identifies each property in the one set with a property in the other, (iii) the structures in the two sets are, respectively, theoretical superstructures and empirical substructures of the models of a theory, and the mapping between them is a general philosophical account of how the theoretical superstructures of a theory's models relate to its empirical substructures.

4. ESSAYS 3 AND 4

A little preamble will be useful before introducing the topics of the third and fourth essays. I said that a physical theory is an ordered pair <M, P> consisting of sets of models representing materialistically and physically possible worlds, respectively, and that the ontology of a theory is the set of simple individuals, basic quantities whose values fix their intrinsic characteristics, and external relations they bear one another.[16] The constitution of the set of materialistically possible worlds relates to the internal structure of the individual worlds in both sets (the latter, in general, being a subset of the former) as follows:

> I) There is a materialistically possible world for every mathematically describable distribution of quantities over, and specification of relations between, basic individuals.

This expresses a sort of semantic truth that is clearest if we think of quantities as dimensions along which individuals can differ

---

14. The transformations in question must be understood as acting simultaneously on elements from both sets, that is, as mapping ordered pairs <Ai,Bj> consisting of an element from the first and second set, respectively, onto other such pairs. In the cases we are most concerned with, the mappings are provided by physical laws, the elements in both sets are physical states represented by equivalence classes of manifolds, and the applicable transformations are just manifold automorphisms.
15. The causal terminology is ill-advised because it narrows the scope of the principle and invites needless controversy.
16. Relations are external if they do not supervene on types of the individuals they relate; unless otherwise indicated, when I speak of relations, I mean external relations.

Introduction

from one another, and n-place relations as quantities pertaining to n-tuples; it just says is that every possible difference in the actual world is an actual difference in some possible world. It has an important consequence, for it entails that all and only transformations which are *not* symmetries of all materialistically possible worlds, correspond to permutations of the value of some quantity or relation pertaining to the actual world, i.e.,

> II) There exists a quantity or relation pertaining to the actual world not invariant under a transformation h *iff* there is some materialistically possible world W of which h is not a symmetry, (i.e., there is a materialistically possible world, W, such that W≠hW).

This in turn means that criteria for distinguishing materialistically possible worlds are also criteria for determining the quantities and relations pertaining to the actual world, and we can approach the latter question by way of the former.

Such an approach would be useful if we had some handle on criteria for distinguishing materialistically possible worlds, and I think that we do.[17] First, if W is observationally distinguishable from hW, then W≠hW. Second, if there are physical laws which are not symmetric with respect to h, then there is a world W, such that W≠hW. The rationale for the first is obvious; observable differences make for physical differences. A more elaborate discussion and some qualifications are included in essay 5. The second is a little less obvious. The way to see it is to appreciate that if the A-determining laws are not h-symmetric, then there is an h-symmetric quantity which is dynamically relevant to A (this is a consequence of Curie's Principle), and if there is an h-symmetric quantity which is dynamically relevant to A, clearly there is a world W, such that W≠hW. Combining these with (II) we get

> (III) There is a quantity or relation which is not invariant under h if *either* there is a world W which is observationally distinguishable from hW *or* if there are laws which are not symmetric with respect to h.[18]

We can strengthen the '*if*' in (III) to an '*iff*' if we adopt a methodological policy of conservativeness in recognizing hidden

---

17. Another reason for preferring the indirect approach is that choices about which transformations are asymmetries of materialistically possible worlds are made un-self-consciously in the course of theorizing, answering only to scientific concerns, and hence are much less likely than questions phrased directly in ontological terms to be guided (tacitly or explicitly) by philosophical presuppositions.

quantities, if, that is, we recognize only quantities we have (observational) reasons to recognize, quantities which are, or which are dynamically relevant to, observable ones.

Now, if we had some way of distinguishing law-governed from chancy phenomena, i.e. of coming up with a list of quantities whose values are determined by laws, we would be in clover, for we could combine it with all of this to obtain a theory-independent guide for deciding what kinds of quantities and relations characterize the actual world.[19] There is a very simple criterion which suggests itself immediately; if correlations are regarded as symptomatic of the operation of laws, then only quantities with a random (or close to random) distribution are chancy. Such a criterion is routinely employed in science, scientists are ever attempting to formulate laws from which all actual correlations between the values of distinct quantities can be obtained from initial conditions which themselves have a random distribution.[20] Bringing all of this together, we can say in a rough and ready way how decisions about physical ontology are made:

---

18. One might suspect the former is a special case of the latter by reasoning as follows: if there is a world M such that M and hM are observationally distinguishable, then h must permute the value of some observable quantity A, and if h permutes the value of A, then surely the A-determining dynamical laws cannot be h-symmetric.
    This reasoning fails because if A is an observable quantity that has more than one value, then there are observationally distinguishable models of the A-determining laws (those in which A takes different values), and if there are observationally distinguishable models of the A-determining laws, then there are transformations which are symmetries of the A-determining laws but which are not symmetries of any of the individual models of the A-determining laws (the automorphisms of the set of A-models which map each model onto a distinct one). So, it's not in general true that if there is a world M such that M and hM are observationally distinguishable, there is an observable quantity A such that the A-determining dynamical laws are not symmetric with respect to h, and both clauses of the disjunction are needed.
19. The law-governed features of the world are just the set of quantities whose values are determined by laws, including probabilistic laws, laws which don't determine the value of a quantity, but entail a probability distribution over its possible values. In this case we can either regard the probabilities themselves as values of a quantity (the 'chance of B') or we can regard them as determining the value of a quantity B* pertaining to large ensembles of systems and whose values are related to the frequencies with systems in the ensemble take one or another B-value.
20. I use 'initial conditions' in a wider sense than is customary, so that it applies to all features of the manifold that are not determined by laws, including the outcomes of objectively chancy events. The rationale for this guide for distinguishing law-governed from chancy phenomena is discussed in the fourth essay, and examples are given that show (this version of) it at work: it

(i) we begin by charting as much as possible of the observable structure of the actual world, i.e. of the actual distribution of observable quantities,
(ii) we then go about trying to formulate laws in such a way that all correlations between observable quantities can be deduced therefrom without assuming overly improbable correlations in initial conditions,[21]
(iii) and these are used to determine via (**III**) what kinds of physical quantities and relations there are.[22]

This is, as I said, rough and ready, but it brings out one important thing: the way in which our physical ontologies are determined by which phenomena we take to be law-governed and the asymmetries of the laws which govern them. In application to non-geometric transformations, it just expresses in a reasonably precise way the familiar methodological injunctions not to rest until all significant correlations in the phenomena have been deduced from laws, and to recognize only quantities which are either observable or dynamically relevant to observable ones.[23]

I hope it is clear even from this crude characterization that there is nothing like a finite effective procedure for fulfilling the

---

is a mark against a theory if it needs to postulate correlations in the initial conditions in of the model which represents the actual world. The more pervasive the correlations, the blacker the mark.

21. I try to say something in essay 5 about the source of the familiar constraints on the form of laws (non-ad hocness, universality, etc.).

22. We can tailor the characterization to apply not just to physics, but to the special sciences. A special science is associated with a set of dependent variables, parameters whose evolution it is concerned to describe and predict. These need not correspond to observable quantities nor need they include all observable properties, the dependent variables of economics, for example, include production rates and inflation. Given any set of dependent variables, the above account functions as a methodological guide for the introduction of independent variables, parameters that are determined to be dynamically relevant to the dependent variables, that need to be introduced to formulate laws with any degree of predictive power.
I should also point out that I have been speaking as though we need to think of our physical theories as providing descriptions of the world, so that the dependent variables of physics include all observable quantities and all organized observable behavior should be derived from the physical laws, but nothing I said depends on this. Whatever the dependent variables of physics, the above provide methodological criteria for introduction of independent ones. Nor need we suppose the dependent variables are known in advance. It might be that we start out with some set of especially important or salient parameters, and let our aspirations be determined by our view of the possibilities. We might, that is, be willing to include whatever variables we find ourselves in a position to harness in the course of theorizing.

theoretical task as described, for our understanding of it changes as our opinions about the phenomenal structure of the world change. We are always collecting data; as we look in new places, guided in part by theory, new correlations emerge while others are found to be mediated by more basic correlations or to obtain only under specialized conditions. Our appreciation of the regularity implicit in the phenomena becomes ever more subtle, and with it the set of correlations which we hold our laws responsible to determining.[24] There is a deeper reason, which will emerge in discussion of the fifth essay, as well, but is too far afield to take up here.

A word about the status of these various principles: **(I)-(III)**: describe the relations of interdefinability between the set of physical quantities and relations, the constitution of the set of materialistically possible worlds, the asymmetries of individual worlds in the set, and the asymmetries of the laws which govern the distribution of quantities. We have no *direct* empirical grip on any one of these sets: there is no series of experiments we could perform which provide a decision procedure for whether an arbitrary world W_1 is materialistically possible, no set of phenomena which would suffice to establish that $(q1...qn)$ are the full set of quantities, no direct empirical test for deciding whether L is a law or h a symmetry of the laws.[25] What we *do* have is an *in*direct empirical grip on some; specifically, we have methodological rules which take us from descriptions of phenomena to guesses about which phenomena are law-governed and what the asymmetries of the laws gov-

---

23. Unobservable quantities/relations not dynamically relevant to systematic differences in observable behavior are not recognized, but we needn't dogmatically deny their existence. The prudent thing is to remain agnostic about them for lack of either evidence or reason to venture an opinion.
24. To make a point inspired by Nancy Cartwright, our theories actually teach us how to create new sorts of orchestrated motion, how to construct what she calls 'nomological machines', systems which exhibit extremely regular behaviors, by shielding out uncontrollable influences.
25. Let me preempt the suggestion that there is at least a negative empirical test for determining whether a transformation h is a symmetry of the laws determining the value of some quantity, viz. if we can produce or find a pair of actual systems related by h which have different values for a quantity B, then the B-determining laws - if such there be - are not h-symmetric. (This covers geometric symmetries as well if we are willing to stretch our terminology so that trajectories count as temporally extended systems, and geometric transformations as transformations pertaining to such systems). The reason this is not a direct empirical test is that it depends on knowing that the pair of systems in question are indeed obtained from one another by h, and this is something which depends on theoretical knowledge. The possibility that any pair of systems we actually have in hand are distinguished by the value of a hidden h-invariant quantity, can only be ruled out if we know what all of the h-invariant quantities are and can

Introduction 15

erning them might be, and these give us a way into the whole circle of notions. The methodological rules are uncontroversial in the sense that they are routinely employed, even if not always explicitly stated. They are commonly enough discussed in application to simple quantities, but almost never in application to spatiotemporal structures, because the dynamical role that spatiotemporal structure plays in our theories is not well understood, having been for so long obscured by the ugly technical apparatus of coordinate systems and coordinate transformations.[26]

I have been discussing how these principles operate in deciding questions about whether certain transformations correspond to permutations of elements of our physical ontology, the quantities and relations which determine the internal structure of the actual world. In the third and fourth essay they are put to work to help us see our way through a pair of old, parallel controversies. The third essay takes its departure from the question of whether reflection is a symmetry of physical *space*, whether any materialistically possible world is identical to its image under spatial reflection. It is one that has loomed large in the philosophical literature on the metaphysics of space since Kant used the contention that a world consisting of a left hand differs from one consisting of its right reflected counterpart as a premise in an argument for the substantivality of space, spawning a controversy that has raged since. The fourth essay takes its departure from a parallel question with a very

---

check that their values are shared. But this is just the sort of knowledge that we were hoping to apply the test to obtain.

This is not to say that it cannot be applied in specific cases, by employing some bits of theoretical knowledge to obtain others if we are careful about how we do it. In cases involving geometric transformations corresponding to physical operations, for example (rotating a system through some angle or setting it into motion), we frequently rely on our confidence that subjecting a system to the physical operation in question changes leaves its intrinsic properties unaffected - whatever they may be. We test the hypothesis that the A-determining laws, say, are h-symmetric by h-ing systems and seeing whether there is an affect on their A-values. The case of the failure of parity in the weak nuclear reaction, discussed in the third essay, provides another nice example in which a very little bit of theoretical knowledge combined with a small set of experimental phenomena - the result a single repeatable experiment - was enough to lead to the rejection of a cherished symmetry of space. The case of thermodynamic phenomena, discussed in the fourth essay, stands in marked contrast by providing an example of a case in which the phenomena point unanimously to the rejection of a certain symmetry of time, but the suggestion is overridden by quite involved theoretical considerations.

26. For discussions of such principles in application to simple quantities, see the literature concerning hidden variables in quantum mechanics.

different history, *viz.* that of whether reflection is a symmetry of *time*. In the latter case, the discussion has focused less on whether we can observationally distinguish a world from its time-reflected counterpart than on whether we need to postulate intrinsic differences between the forward and backward directions in time in formulating dynamical laws. Both discussions benefit from being seen against the background of the general principles while, I hope, lending support by showing them persuasively at work.

5. ESSAY 5

Thus the first four essays. I indicated something of the content of the fifth in the description of science with which I started. It is the philosophical heart of the thesis, an attempt to spell out in the kind of detail that is needed to see if it will work, the simple idea that science is about trying to map the structure of the world on the basis of experience. It begins by noticing the spatiotemporal and sensory limitations on experience. Even supposing, as we must if we are to generalize from our experience at all, that the distribution of qualities in our region of the universe is characteristic of its distribution throughout, the accessible information includes only information about the distribution of qualities in the region which we have already experienced and the (relatively) local, invariant relations between these which can be discerned and (granted, again, the assumption that they are characteristically distributed) projected into regions that presently lie outside our experiential ken.[27] I suggested that we understand the process of coming up with a physical theory as a matter of seeing how to shift back and forth between world descriptions in terms of different sets of quantities with an eye to fixing on one which best fulfills the following desiderata: it can be determined from knowledge of the distribution of qualities, qualitative structure supervenes on it, and it is dominated by the local, invariant relations that make the most powerful basis from which to project.[28]

---

27. This is a little strong; the rejection of skepticism about simple induction actually only requires some principle relating the distribution of qualities in the observed region of the universe to its distribution throughout. The most natural such assumption, and the one that we instinctively employ, is that the former is characteristic of the latter.

So, as it is depicted in the fifth essay, theorizing is largely a matter of choosing a set of quantities in terms of which to describe the world and an important component of the choice is deciding just which quantities there are. The second, third, and fourth essays, to the extent that they are concerned with the principles which guide this decision, fill in parts of the bigger picture painted in the fifth.

I have concentrated in the latter on spelling out a story about science in a tight way, without undue rigor and shorn of extraneous detail. Though it will be obvious to anyone immersed in the literature that it was worked up partly in the image of stories other people have told, partly as a reaction to problems they encountered and objections raised against them, I have done little in the discussion to relate it to any of these, because I don't know how to do that without getting bogged down and losing the main point in erudition. The hope is that the terrain is well-enough known (which paths are dead ends, which are blocked by insurmountable obstacles, where lay mines...) that the uglier parts of the path I have taken through it, the detours and the departures from straightness, don't look completely arbitrary.

---

28. I have left out some niceties which are taken up in the fifth essay.

Essay 1

Symmetry: An Elementary Introduction

1. INTRODUCTION

Symmetry has a chameleon-like role in physics. For every type of structure we encounter there is a corresponding type of symmetry: geometric symmetries of particular physical objects, symmetries characteristic of a given type of object, symmetries of abstract objects like states, symmetries of state-*spaces*, ... and so on. I want here to introduce the general notion of symmetry and illustrate it by applying it to some especially simple and familiar sorts of structures. None of the details are important; the reason for taking up these structures in particular is that it is to them that the notion of symmetry has its most immediate and easily visualizable application, and the abstractions which follow on its generalization to the kinds of structures that arise in physics will go more smoothly if we have them clearly in mind.

2. GEOMETRIC SYMMETRIES:

Some preliminary definitions: a *mapping* **S** of space-time associates with every point **p** a point **p'** as its image. One such mapping is the identity **I** which associates every point with itself. Another is rotation about a point **O** by $n°$ which takes every point **a** into the point **a'** such that the angle **aOa'** is $n°$, and yet another is translation through a distance d in a particular direction which associates every point with one separated from it by d. Given two mappings **S** and **T**, where **S** carries **p** into **p'** and **T** carries **p'** into **p"**, the *composition* of **S** and **T**, written **ST**, is a mapping which carries **p** into **p"**. Composition is not in general commutative; if **S** is a horizontal translation which carries c, below, into c1 and **T** is a rotation through $90°$ about c, for example, **ST** will take c onto c_2 whereas **TS** will take c onto c_1.

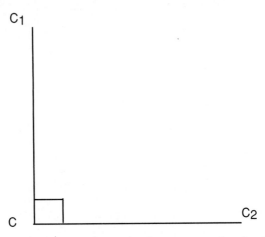

Reflection in a plane through a line L is another kind of mapping which associates with every point another the same distance from L on the opposite side. *Transformations* are one-one mappings, which is to say that they associate any point with only one other. If **S** is a transformation which carries **p** into **p'**, then **S**'s *inverse*, S^{-1}, carries **p'** into **p**. Transformations, unlike mappings, always have inverses, and it should be obvious that the composition of a transformation and its inverse is the Identity, for if the first transformation takes you away from your starting point, the second will bring you back; it is the job of **S**'s inverse to undo **S**'s work: $SS^{-1} = S^{-1}S = I$. Just as obviously, the Identity is its own inverse, as is reflection, and performing S then T is always the same as performing the inverse of **T** and then the inverse of **S**; $ST = T^{-1}S^{-1}$.

Symmetry: An Elementary Introduction

The symmetries of a structure are the transformations which map it onto itself. The Identity is a symmetry of all structures, and if **S** is a symmetry, **S**$^{-1}$ is also. Moreover, and if **S** and **T** are symmetries, so is **ST**. This means that the set of symmetries of a given structure form a group, which is to say it is closed under inverse and composition and contains the Identity.

A space is a structure, as are the figures embedded in it, and the symmetry group of a figure is clearly a subgroup of that of the space in which it is embedded. The structures whose symmetries I want to look at are Euclidean figures in two and three dimensions, all those shapes that can be drawn on a piece of paper or constructed out cardboard or concrete, if we suppose for simplicity that physical space has a Euclidean structure.[1]

The symmetries of Euclidean space, 'congruences', preserve all distances between points, and they include transformations which interchange the left and right sides, as well as those which don't ('improper' and 'proper congruences', respectively; the latter are also called 'motions' because they preserve shape and scale and correspond to the action of carrying a body from one part of space to another). The congruences and the proper congruences both form groups, but the improper congruences do not because the set of them is not closed under composition (performing one improper congruence after another gives you a proper congruence, not an improper one). The proper congruences are said to form a subgroup of congruences of index 2, which means that all improper congruences can be obtained from any one, by composing it with each of the proper congruences. This is the sense in which the proper and improper congruences each make up half of the full group of congruences.

The symmetry group of a figure, as I said, is just that subset of the symmetries of the space in which it is embedded that map it onto itself. So, for example, the circle in a plane with center O, on the left has the symmetry described by the continuous group of all plane rotations. The pentagram is less symmetric, it is carried into itself only by the five proper rotations around its center whose angles are multiples of $360°/_5$, the identity, and the five reflections

---

1. Not just figures which can be drawn by us with the amount of time, paper, patience, etc. that we actually possess, but all those that can be drawn by an infinitely talented, infinitely patient being, with an infinite amount of paper and time.

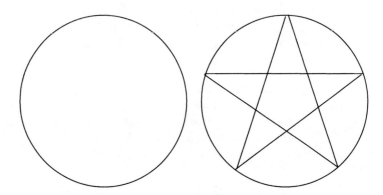

in the lines through the center and each of the vertices. All of these are symmetries of Euclidean space.

Symmetry with respect to reflection ('bilateral symmetry') is probably the most ubiquitous symmetry both in art and in nature, at least among land animals. A two-dimensional figure is bilateral-

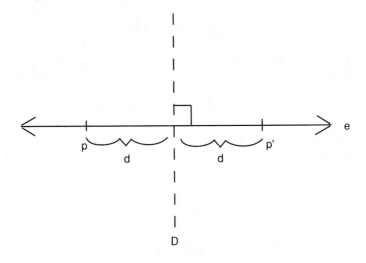

ly symmetric with respect to line D if it is carried into itself by reflection in D.

Take any line l perpendicular to D and any point **p** on l, there exists exactly one point on the other side of D on l which has the same distance from D as **p** (it will be identical to **p** only in the case that **p** is on D). This point is 'mirror image' of **p** with respect to

D, and reflection through D maps each point onto its mirror image in D. In three dimensions we say a figure is bilaterally symmetric with respect to a *plane* D' just in case it is carried over into itself by the mapping which associates with an arbitrary point **p** its mirror image on the line perpendicular to D'. If the human body were perfectly bilaterally symmetric, every curve, every mole and crease on one side would have a perfect replica in the corresponding position on the other, so that if a person were sliced down the navel and a mirror placed along the cut, the image in the mirror would be a perfect representation of the absent half. Most ancient Sumerian and Egyptian ornamental representations of people approaches bilateral symmetry, as does some Greek sculpture, but in reality, of course, such bilateral symmetry as humans possess is only skin deep, and is never perfect.[2]

3. FINITE GROUPS IN ONE AND TWO DIMENSIONS

Let's look more systematically at some of the groups these symmetries form. In one dimension, the only possible improper congruences are reflections, and the only proper congruences are translations. A figure invariant under a translation **T** (recall that translations map every point onto one at some fixed distance in a certain direction), shows repetition in a regular spatial rhythm and is invariant also under the iterations of **T**, as well as the identity, the inverse of **T**, and the iterations of the inverse. If **T** shifts the line by a distance a, then the nth iteration of **T** shifts it by na (n=0, + or -1, + or -2...). This means that all groups of translations must be infinite and that all of the transformations in a given group must be (in this sense) multiples of a single basic translation **a**. One-dimensional examples are easy to find: repetition of a sound at equal intervals - in musical parlance, rhythm - is a translatory symmetry in time. Besides translations, reflections are possible in one

---

2. It has often been said that a large component of human beauty consists in bilateral symmetry of features. The fact that beauty is often felt most deeply in idiosyncratic departures from symmetry supports rather than (as one might suppose, at first) tells against the suggestion, for, it is not a new observation that the most affecting departures from symmetry occur against a symmetric background and have the effect of highlighting the overall symmetry rather than detracting from it. Elizabeth Taylor's mole is something on Elizabeth Taylor's face that it wouldn't be on Barbara Streisand's, just as a preponderance of dark on one side of a painting is more arresting if the painting is organized around bilateral symmetry than it would be otherwise.

dimension. To take another example from music; reflection is often used in the construction of fugues, though it is harder to recognize than rhythm. Reflective and translatory symmetries can be combined in particular figures; a reflection followed by the translation through OA yields reflection in that point which halves the length of OA. The best spatial examples come from ornamental art where repeating, symmetric patterns are often used to fill long bands of space. In general, finite symmetry groups in one dimension contain only the identity and reflections.

When we go from one dimension to two, we have to add to the translations and reflections, and this opens up the possibility of rotational symmetries. I mentioned earlier some examples of finite two-dimensional figures that are bilaterally symmetric, as well as the examples of the circle and pentagram which combined rotational and reflexive symmetry. With these in hand, it is easy to come up with an exhaustive list of finite groups of both improper and proper rotations in two dimensions. When we ascend further to three dimensions the list has to be extended again, but only slightly. I said that any group of translations of a line, provided only that it contains no transformations arbitrarily near to the identity except the identity itself, consists of the iterations of a single translation. One consequence of this is that any finite group of proper rotations around a point O in a plane (or around a given axis in space) contains a smallest rotation T whose angle is an aliquot part $a = 360^0/n$ of the rotation by 3600, such that the group consists of the iterations of T: $T^1, T^2, T^3, \ldots T^n = I$. To see this, take a line or a band exhibiting this type of symmetry where the repeated section has length a, and sling it in a circle with O at its center with circumference an integral multiple of a, e.g., $12a$. The pattern obtained will be carried into itself by the rotation around O through $a = 360^0/{12}$, together its repetitions. The twelfth iteration will be the identity, a complete rotation through 360^0, the thirteenth will be a rotation through a again and hence identical to the first, and so on. The new pattern determines a finite group of rotations of order 12, i.e. one consisting of 12 operations, and the order n completely characterizes the group. The set of groups so obtained itself has a group structure and is called the cyclic group, written $C_n = \{C_1, C_2, C_3 \ldots\}$.

Groups of *improper* rotations are obtained as follows: let Z be a reflection in O, that is, let it carry any point p into its antipode p' with respect to O found by joining p with O and prolonging the straight line from p to O by its own length so that pO=Op'. Z commutes with every rotation S, i.e. SZ=ZS. Now let C_n be one of our finite groups of proper rotations, add to the rotations S in C_n all improper rotations of the form ZS. The result is a group of order 2n (i.e. twice that of C_n), consisting of the repetitions of a single improper rotation by an aliquot part $a=360^0/$ of 360^0, together with its iterations and inverses, and each of these combined with a reflection in n axes forming angles of $1/2a$. This set of groups is the dihedral group, $D_n = \{D_1, D_2, D_3 ...\}$. A figure whose group is C1 has no symmetry at all, while one invariant only under D1 has bilateral symmetry. Together these exhaust the possible groups of finite rotational symmetries in two dimensions.

Examples of physical objects with various of these symmetry groups are easy to find: in the organic world, flowers are the most obvious, and their symmetry groups, most commonly of orders 3 or 5. In the inorganic world, crystals obviously exhibit the most perfect rotational symmetries, but, for reasons having to do with their microscopic structure, only those of orders 2, 3, 4 and 6 are physically possible. Rotational symmetries are very uncommon among land animals, but some of the lower sea animals exhibit them to an remarkable degree.[3]

Moving away from concrete examples and back to mathematics, we have seen that all finite groups of automorphisms consist entirely of congruences, but if widen our gaze to include infinite groups, we have to consider one more type of transformation: dilations. Intuitively, a dilation preserves the shape of a figure while allowing its scale to change; formally, it has a fixed point O and carries each point into one a ratio a:1 of the distance of the original from O, and iteration of the dilation S generates the group of dilations: $S_n=\{S_0, S_1, S_{-1}, S_2...\}$. These three simple families, the improper and proper rotations (C_n and D_n, respectively), and the dilations, (S_n), exhaust groups of symmetries in two dimensions.

---

3. It is not hard to see why; even if the philogenetic laws are invariant under rotation, the factors which influence their development are not. The direction of gravity, in particular, is an important factor, and narrows the set of symmetries from all rotations around the center p to all rotations about an axis, and the direction in which an animal moves - whether in water, air, or on land - is another source of asymmetry. After determination of the front-back (antero-posterior) and top-bottom (dorso-ventral) directions, only the distinction between left and right remains, and hence only bilateral symmetry is to be expected.

We can think of their relations to one another in the following way: the minute hands of some watches rotate in a continuous uniform way around the face of the watch, while others jump from minute to minute. The continuous motion of the former are regarded as just what happens when we let the intervals between which the hand jumps become smaller and smaller until they are infinitesimal. Just as a full rotation of the minute hand of the latter watch is the result of 60 repetitions - one each second - of a rotation through an angle $360°/60$, the continuous motion of the former consists of the infinite repetition of the same infinitesimal motion in consecutive infinitely small time intervals of equal length. The former form a discontinuous group (C_{60}) which is naturally regarded - along with the group formed by taking any rotation by an integral number $360°/_n$ and its iterations and their inverses - *as contained* in the continuous group. Similarly, we regard the finite groups of translations as embedded in the continuous group representing the uniform motion of a space-filling substance. The set of transformations $U(t, t')$, each of which carries the position p(t) of any point of the substance at time t over into its position p(t') at t', form a one-parameter group so long as $U(t, t')$ depends only on the interval between t and t', i.e. so long as the fluid is in uniform motion, $U(t, t')=S(t'-t)$. The group law $S(t1)S(t2)=S(t1+t2)$ ensures that the motions during two consecutive time intervals t_1, t_2 result in the motion during the interval t_1+t_2, the motion during one minute leads to a definite transformation $S=S(1)$, and for all integers n the motion $S(n)$ during n minutes is the iteration S^n. In this way the discontinuous group consisting of the iterations of S is embedded in the continuous group with the parameter t consisting of the motions $S(t)$.

The same goes for transformations which are neither simple rotations nor mere translations: any such transformation has a fixed point O, consists of a rotation about O combined with a dilation from the center O, and can be obtained as the stage reached after 1 minute by a continuous process **S(t)** of some combined uniform rotation and expansion. This process carries a point (any point but O itself, that is) along a logarithmic (or equiangular) spiral which has the property of going over into itself by a continuous group of automorphisms. The straight line and circle are the only two dimensional figures which have this property, and they arise from this curve when the former and latter components in the rotation-plus-dilation, respectively, is Identity.

4. RECAP AND GENERALIZATION

Thus the walking tour through groups of symmetries of two-dimensional Euclidean figures. We have seen that the set of transformations which are the symmetries of any figure has a group structure and any group of transformations is the symmetry group of some figure. It follows from this that a complete list or catalogue of groups of transformations in two and three-dimensional Euclidean space translates into a complete catalogue of the groups of symmetries possible for two and three dimensional Euclidean figures. These, as we saw, fall into three very simple families, and so it turns out that the symmetry groups possible for the whole vista of two and three-dimensional Euclidean figures - anything that can be written down on a piece of paper or constructed out of cardboard or concrete - can be taken in just by grasping the transformations in these families. Moreover—and this is the really important reason to study structures from the point of view of their symmetries—it imposes an organization on a set of structures, in terms of shared elements in their corresponding groups, that turns out to be an extremely fruitful way of categorizing them. Not only does it capture intuitive judgments of relative similarity (a structure A is more similar to B than C is if the set of transformations C shares with B is a proper subset of those A shares with B), but it captures many of the relations between structures that turn out to be important in physical contexts. These are claims that, hopefully, will be illustrated in the essays to follow.

---

4. The notion of structure is most commonly introduced indirectly by defining isomorphism, i.e. sameness of structure. This is how Russell does it in Introduction to Mathematical Philosophy, Principles of Mathematics, the introduction to Principia, and Human Knowledge. Here, for example, is what he says in the last:
"We can now proceed to the formal definition of 'structure'...Out of the terms of a given class many structures can be made... Let us in the first instance confine ourselves to dyadic relations. We shall say that a class a ordered by the relation R has the same structure as a class b ordered by the relation S, if to every term in a some one term in b corresponds, and vice versa, and if when two terms in a have the relation R, then the corresponding terms in b have the relation S, and vice versa ... the definition of identity of structure is exactly the same for relations of higher order as it is for dyadic relations." (p. 254).

The generalization of the notion of symmetry to non-geometric contexts is straightforward; a structure, as I said, is any set of elements with a relation defined over them, and the symmetries of a structure are those transformations which map it onto itself.[4] We can confine our attention to relations pertaining to the basic elements since all others will supervene on these, and it is convenient to separate the one-place relations which characterize the basic elements internally from the external relations between them. Any structure can be completely described by specifying the number of its basic elements, their types, and the network of external relations they bear one another.[5] The geometric figures we have been talking about consist of an infinite set of intrinsically undifferentiated points related by a network of distances. Likewise, non-geometric mathematical structures; the structure of the set of rational numbers, for instance, is that of a denumerable infinity of intrinsically undifferentiated, algebraically related elements.[6] The manifolds which I described in the introduction as forming the geometric background for space-time theories, consist of a nondenumerable set of points (or events), intrinsically differentiated by the values of the field quantities instantiated at them, related externally by locally Euclidean network of spatiotemporal relations.

Corresponding to the distinction between intrinsic properties and external relations is a distinction between two types of transformation: those which consist of exchanges between different types of basic element, and those which work on external relations. I refer to these, respectively, as **geometric** and **non-geometric transformations**, stretching the terminology a bit beyond its usual scope, but not in a way that should be confusing. It should be clear now that, where I wrote above that a structure is

---

5. An external relation is one which doesn't supervene on the intrinsic properties of its relata; a purely external n-place relation is one which varies completely independently of the intrinsic properties of its relata, one which any n elements may bear one another, regardless of their intrinsic properties. The intrinsic natures of and purely external relations between a set of elements will fix the external relations between them which are not purely external, and it can be shown that there will always be a single, purely external, two-place relation which fixes the whole network of purely external relations between basic elements. So to completely describe a structure, we really need only specify the number and types of basic elements and the purely external two-place relations they bear one another.
6. Algebraic relations are just those that can be obtained from the four basic operations +, -, x, and ÷.

completely characterized by the number and types of its basic elements together with the network of external relations they bear one another, I could just as well have written that a structure is completely characterized by its geometric and non-geometric symmetries. For, to know these is to know which elements and ordered sets of elements can be substituted for one another without affect (i.e. which elements are of the same intrinsic type and which sets of them bear one another the same external relations, respectively), and to know that is to know everything there is to know about a structure. In the penultimate paragraph of his booklet *Symmetry*, Herman Weyl writes:

> "What we learn…and what has indeed become a guiding principle in modern mathematics is this lesson: Whenever you have to do with a structure-endowed entity, try to determine its [symmetries], the group of those element-wise transformations which leave all structural relations undisturbed. You can expect to gain a deep insight into the constitution of it in this way." (p. 144)

Take this, and supplement it with the suggestion that science (or at least a good and important part of it) is engaged in trying to map the structure of the physical world, and you have an excellent motto for the essays which follow.

Essay 2
Curie's Principle

> As far as I see, all a priori statements in physics have their origin in symmetry.
>
> (Weyl, 1952, p. 126)

1. INTRODUCTION

In 1894 Pierre Curie published a paper in which he stated the principle that the symmetry of a cause is always preserved in its effects. The proof of the principle is simple, yet far from being recognized as an important mathematical truth with wide-ranging applications in physics, when it receives any discussion at all, it is either dismissed out of hand or allotted the status of a mere methodological guide. Even as a methodological guide it is alleged to apply only in deterministic contexts, and hence to hold little interest for today's physics. Here are some pronouncements that are typical of the literature.

> Apart from being logically incorrect, [Curie's Principle] is in obvious contradiction with empirical evidence. (Radicati 1987, 202)

> What is to be said of this fundamental, profound principle that an asymmetry can only come from an asymmetry? The first reply is that qua general principle it is most likely false and certainly untenable. (van Fraassen 1989, 240)

> Curie's putative principle (even in my formulation, which did not use causal terms) has no fundamental ontological status... it betokens only a

thirst for hidden variables, for hidden structure that will explain, will answer why?— and nature may simply reject the question. (van Fraassen 1991, 24)

Grounds given for rejecting the principle are typically one or more of the following:

(i) the principle purports to be *a priori* but actually rules on an empirical question;
(ii) the principle is simply false, since nowadays physicists recognize many phenomena which spontaneously break the symmetry of the preceding conditions;
(iii) the principle—even if it is true for deterministic theories—has no application in indeterministic contexts; and finally,
(iv) the principle, if interpreted in such a way as to make it true, is empty.

I will argue *contra omnes murmurantes*:

(i) that the principle, properly understood, is necessarily true;
(ii) that instances of (so-called) 'spontaneous symmetry breaking' are not counterexamples to it;
(iii) that even in the context of indeterminism, it remains a powerful heuristic; and finally,
(iv) that it has important and far-reaching consequences in physics and in the philosophy of science quite generally.

Recognizing the truth of the principle requires properly understanding it, and this in turn requires appreciating Curie's basic insight into the physical significance of the symmetries of a set of laws and the states they relate. It is an insight well worth appreciating, for the physical and philosophical rewards are great, and some of these will be indicated in what follows. Let's turn to the principle to see what all of the fuss is about.

2. THE PRINCIPLE

A physical theory specifies what sorts of objects there are and how they are related to one another, so theories in physics—at least in part—describe structures. Any structure uniquely determines a group of transformations called its symmetry group, and the study of such groups is the mathematical theory of symmetry. It is a particularly elegant theory in that the physically interesting relations between structures receive an exceedingly simple expression in terms of relations between their symmetry groups, and all of the important particular truths about structures can be derived from simple principles relating their symmetries. Curie's Principle is such a principle. Curie states it several times in his paper departing little from the formulation given in the first paragraph:

> When certain effects show a certain asymmetry, this asymmetry must be found in the causes which gave rise to them. (Curie 1894, 401)

What the principle says depends crucially on how the terms 'cause' and 'effect' are understood, so bear with me through the definitions; exegetical support for the reading will be given in section 9. Let A and B be families $\{A_1, A_2 ...\}$ and $\{B_1, B_2 ...\}$ respectively, of mutually exclusive and jointly exhaustive event types, and let the statement that A is a **Curie-cause** and B its **Curie-effect** mean that the physical laws provide a many-one mapping of A into B, or—more simply—that (relative to the laws) A determines B.[1] A and B may be different aspects of the total state of one system at a single time, or total states of a system at two different times. Curie-causes (and effects) relate to their specifications as physical quantities relate to their values. Consider, for example, the Boyle-Charles law relating the pressure and volume of an ideal gas to its temperature: $PV=kT$. The product of the pressure and volume is, in this instance, the Curie-cause, the temperature is its Curie-effect, and particular values of the two which satisfy the laws are specifications of the same. Similarly, consider the state of an isolated Newtonian system at two different times. The Newtonian dynamical laws are deterministic, so the prior state is the Curie-cause, the later one is its effect, and pairs of particular states which satisfy the equations are specifications. In what follows, I will use subscripts to distinguish Curie-causes and effects from specifications, so $<A,B>$ will denote a Curie-cause and—effect, whereas $<A_i, B_i>$ will denote one of its specifications.

---

[1]. I'll suppress the relativity from here on, though talk of A being the Curie-cause of some effect B is always to be understood as tacitly relativized to the laws of a theory.

Now, let us call the symmetries common to all specifications of some Curie-cause its **characteristic symmetries**, and let us call **idiosyncratic** those transformations that are symmetries of some but not all specifications. In the case of the Boyle-Charles law above, transformations under which all specifications of the values of pressure and volume are invariant, e.g., spatial reflections or permutations of the value of some unrelated parameter, are characteristic symmetries of the Curie-cause. By contrast, transformations under which only some specifications are invariant, are idiosyncratic symmetries of those specifications. Exchange of the values of temperature and pressure (in standard units), for example, is an idiosyncratic symmetry only of states for which the two happen to be equal. Likewise, in the case of the Newtonian dynamical laws, transformations under which all states are invariant, e.g., simple spatial displacements, are characteristic symmetries of the Curie-cause, whereas transformations under which only some states are invariant, e.g., exchange of values of mass and acceleration, are idiosyncratic symmetries of those states. I propose that we interpret 'cause' and 'effect' in Curie's statement of his principle as Curie-cause and Curie-effect. The content of the principle, then, is that all characteristic symmetries of a Curie-cause are also characteristic symmetries of its effect.

Equivalently, if T is an idiosyncratic asymmetry of any specification of a Curie-effect, then it is also an idiosyncratic asymmetry of some specification of the corresponding Curie-cause.[2]

3. THE PROOF

The proof of the principle follows almost immediately from the definitions in the preceding section. The laws of a deterministic theory enable the Curie-effect B to be derived from the cause A, and can be represented as a mapping of the set of possible specifications of A into the set of possible specifications of B, equivalent to a set (usually infinite) of ordered pairs $<A_i, B_i>$ where each $<A_i, B_i>$ is a solution to the laws. If $<A_j, B_j>$ is such a pair and T

---

2. It follows from the definition of a characteristic symmetry that T is an idiosyncratic asymmetry of (i.e. is not a characteristic symmetry of) some specification of a A (or B) iff it is not a characteristic symmetry of A (or B). We get the second formulation of Curie's Principle by taking the contrapositive of the first (i.e., if T is not a characteristic symmetry of a Curie-effect B, then it is not a characteristic symmetry of its Curie-cause, A), and replacing the antecedent and conclusion with their equivalents under the definition of characteristic symmetry.

Curie's Principle

is a transformation which acts on A and B, then T takes $<A_i, B_i>$ onto $<TA_i, TB_i>$. Now, suppose Curie's principle isn't true. Then there is some T, and some pair of solutions of the laws $<A_i, B_i>$ and $<TA_i, TB_i>$ such that $TA_j = A_j$ but $TB_j \neq B_j$.[3] But the laws are deterministic, so any Curie-cause has only one physically possible effect among the B's and it follows—contrary to the hypothesis—that $TB_j = B_j$, and Curie's Principle is true after all. If A is a Curie-cause of B and T is a characteristic symmetry of A (i.e. if it acts as the identity on each specification A_i of A), it had better act as the identity on each Bi as well.

Let me emphasize that the asymmetries in question are characteristic; T is a characteristic symmetry of A *iff* it is a symmetry of each of the A_i's (i.e., *iff* for all A_i, $(TA_i = A_i)$), and it is a characteristic symmetry of B *iff* it is a symmetry of each of the B_i's. If this is not kept in mind, there is a temptation to think that cases like the following provide counterexamples: consider a world consisting of two types of particle, bald and hairy, governed by deterministic dynamical laws which prescribe that all hairy particles decay into bald ones, but bald ones never decay into hairy ones. Take as the Curie-cause partial state-descriptions of the form

<# of hairy particles present, # of bald particles present> and consider the transformation T that replaces all bald particles with hairy ones and hairy particles with bald ones

$$T: <m,n> \longrightarrow <n,m>.$$

Now, the state description $A_1 = <3,3>$ nomologically determines $B1 = <0,6>$, and T is a symmetry of A but not a symmetry of B. This is correct, but it does not constitute a counterexample to Curie's Principle because T is not a *characteristic* symmetry of A, it is merely an idiosyncratic symmetry of A_1. The A_i's are essentially *ordered* pairs, for they are not all invariant under exchange of bald and hairy particles; having 6 bald and 0 hairy particles is quite different from having 0 bald and 6 hairy ones.

It may still seem that I have pulled a rabbit out of a hat: I have purported to show that it is impossible to write down a set of deterministic equations that carries the state of a system charac-

---

3. Notice that the fact that we can write down pairs of specifications <Ai,Bi> and <TAi,TBi>, where TAj=Aj but TBj≠Bj (as, for example, is the case if A and B are far enough apart that T operates on B but not on A) is unimportant. What is ruled out by Curie's Principle is the existence of pairs of solutions to the laws <Ai,Bi> and <TAi,TBi>, such that TAj=Aj but TBj≠Bj.

terized by certain characteristic symmetries onto a state that lacks those symmetries. Even if this is plausible enough with respect to non-geometric transformations like the one above, one might expect geometric transformations to provide counterexamples. For the sake of uniformity in talking about geometric transformations, and for reasons that will be given in section 4, I will restrict attention to the generally covariant formulations of theories and I will assume that the transformations in question are manifold *automorphisms*, i.e. one-one suitably continuous and differentiable mappings of a manifold M into M. A system of equations is covariant under a transformation T just in case for any <A, B> that is a model of the equations, <TA,TB> is a model as well; and a system of equations is **generally covariant** just in case it is covariant under arbitrary manifold automorphisms.[4] No generality is lost because all theories can be given a generally covariant formulation and any transformation can be represented as a manifold automorphism.[5]

Now, imagine a universe which exists for exactly a minute and consists of a sphere which gradually deforms into an ellipse; surely it is possible to write down a deterministic equation describing the evolution of the sphere. Try to do it, however, and you will find that you will need to include a parameter which takes different values for different directions in space, i.e. a parameter whose value is not invariant under arbitrary spatial rotations. You will need to do so because you will need to distinguish the direction along which the sphere elongates. This is precisely to recognize a characteristic asymmetry in the Curie-cause of the elongation of the sphere. Nothing very mysterious is going on here: if A is the Curie-cause of B, then A nomologically determines B. This means that there is a many-one mapping from the set of specifications of A into the set of specifications of B, so different B_i's always (i.e. in all physically possible worlds) 'come from' different A_i's. From

---

4. Here, as above, where a definition is given, the defined term is indicated by bold type.
5. Let me emphasize that the covariance group and the symmetry group of a theory are two different things. The covariance group is the set of coordinate transformations which preserve the truth of the laws. The symmetry group of a theory is the set of automorphisms of its solution set, the transformations which never take you from a solution onto a non-solution, or vice versa.
6. Since it is not necessarily the case that the A's can be represented as a function of the B's (i.e. that the mapping is one-one), it doesn't follow that differences in the A's are always accompanied by differences in the B's, and hence it doesn't follow that the asymmetries of the A's are also asymmetries of the B's. I will come back to this in section 8.

this it follows straight-away that the intrinsic asymmetries of B are also intrinsic asymmetries of A.[6]

4. INTERPRETATION OF THE PRINCIPLE

Let me spell this out in a way that brings out the physical significance of the symmetries of a set of equations: instead of thinking of them as automorphisms of the set of solutions, we should think of them as the set of transformations among the values of relevant parameters which preserve their truth. The symmetries of a set of equations determining one among a family B of alternatives, then, correspond physically to either

(i) permutations of the values of B-irrelevant parameters, or
(ii) irrelevant permutations of B-relevant parameters, i.e. transformations which either map them onto themselves or are accompanied by compensating transformations in the values of other parameters in such a way as to preserve the relation described by the law.

The key to understanding Curie's Principle is to focus on the contrapositive; transformations which *aren't* symmetries correspond physically to relevant permutations of the values of relevant parameters. This is easy to see in the case of non-geometric asymmetries, since these correspond to the transformations of values of parameters in the equation expressing the laws in question, but it has not always been so transparent in the case of geometric symmetries. The problem is that if dynamical theories are formulated in their traditional coordinate-dependent manner and geometric transformations are represented as transformations between coordinate-systems, T may be an asymmetry of the laws determining B, even though no T-asymmetric parameter appears in the B-determining equations. This, combined with the historical confusion about the precise nature of the coordinate-dependence, obscured the physical significance of geometric transformations for generations. It is only in hindsight and by concentrating on their coordinate-*in*dependent (i.e. generally covariant) formulations that it becomes clear that geometric asymmetries of a set of B-determining laws can be understood in precisely the same manner as their non-geometric asymmetries, *viz.* as relevant permutations of the values of B-relevant parameters. And this is because it is only in the generally covariant formulations that the geometric asymmetries of the laws change the value of some parameter in the B-determining equations.

A slightly more detailed discussion is given in the appendix, but a quick example will help to illustrate the kind of confusion encouraged by the coordinate-dependent style of formulation. Consider an isolated Newtonian system consisting of a ball at rest on a frictionless surface. Let A1 be the coordinate description of the state of the ball at a time t_1 and B_1 its state one minute later at t_2 relative to a coordinate system with respect to which the ball is at rest. Suppose that the ball remains isolated in the interim. $<A_1,B_1>$ is a solution to Newton's laws, and since the laws are deterministic, A_1 is the Curie-cause of B_1. Now, consider the coordinate transformation T which takes every point (x,y,z,t) onto the corresponding point $(x,y,z+at^2,t)$. T carries $<A_1,B_1>$ onto $<TA_1,TB_1>$ which has the ball spontaneously accelerating in the z direction with no force acting on it, so although $<A1,B1>$ is a solution to the laws, $<TA_1,TB_1>$ is not. The situation is usually described by saying that Newton's laws hold 'relative to' the first coordinate system but not 'relative to' the second, and hence are not covariant with respect to transformations between the two. A better way to describe it is to say that there is some dynamically relevant difference between $<A_1,B_1>$ and $<TA_1,TB_1>$, a difference which is obscured by the fact that they get the same coordinate-description, simply relative to different coordinate systems. For surely, one is inclined to think, it doesn't matter which coordinate system one describes a given system in terms of. The difference shows up much more clearly on their coordinate-independent representations where there is no relativity to a coordinate system and the difference between $<A_1,B_1>$ and $<TA_1,TB_1>$ is explicitly represented.

This might be a good place to pause and say a word about the relation of Curie-causes and—effects to what we ordinarily call initial and final conditions. On the one hand, we think of the initial conditions relevant to some effect as the set of conditions which—by the lights of our theory—are sufficient to ensure its appearance, in the sense that in all models of the theory in which the initial conditions obtain, the effect obtains also (and which are such that, moreover, this is not true of any proper subset of them). On the other hand, we don't typically include among the initial conditions relevant to an effect B everything that goes into the specification of its Curie-cause, A. In particular, we don't include the intrinsic spatiotemporal structures which distinguish systems related by a geometric transformation that is *not* a symmetry of the laws. With respect to the Newtonian ball above, for example, something which fixes the inertial or non-inertial character of the system's motion must be included in A, but there is nothing in what we usually regard as the relevant initial conditions which does that, so these cannot be identified with the Curie-cause of B.[7] I think this is a real tension in our use of the notion of initial conditions. Whether we want to revise it by giving up the gloss, or by extending the notion of initial conditions so that they include everything that goes into the specification of the Curie-cause, is up for grabs.

5. HOW ONE COULD DOUBT THE PRINCIPLE

Curie's Principle appears so simple and obvious on the reading I have given that some explanation is needed for why people have doubted it. There are a lot of ways one might go wrong: one reason, surely, is the general confusion surrounding the physical significance of geometric transformations, but there are others. Among them is the fact that Curie seems to have assumed that the world is governed by deterministic laws and so he often speaks as though we can infer simply from the existence of an asymmetry in the state of a system at a time, that there was an asymmetry in the state which preceded or that asymmetric causes intervened from the outside. This is a mistake; the world might be—and according to our current theories *is*—riddled throughout with indeterministic events. We should keep the assumption of determinism separate from Curie's Principle and allow for indeterminism by reformulating the principle to say that we can infer from the existence of the asymmetry in some individual effect *either* that there was an asymmetry in the Curie-cause *or* that there was no Curie-cause. This is not to suggest, however, that the principle is useless in indeterministic contexts; far from it, but I will leave its application in such contexts to section 6.

---

[7]. We usually regard the positions and momenta of the particles which constitute the ball as the relevant initial conditions, without any specification of whether the coordinate system with respect to which they are given, is an inertial one.

Another way to doubt the principle is to restrict what counts as a cause in such a way as to rule out some of the conditions which go into the specification of the Curie-cause. Return again to the Newtonian ball of the last section; I said that something which fixes whether the system is moving inertially or accelerating must be included in the Curie-cause. One might insist that no such thing could be included in the *cause* because the state of motion of a system is not one of its intrinsic features and only intrinsic features can be counted among the causes of its behavior. I don't think there is much to be said for such a position; it is clear from the examples Curie uses in his paper to illustrate the principle that he had no such restricted notion of cause in mind, nor is it consistent with our usage in either scientific or non-scientific contexts. We count the speed at which the car was traveling among the causes of the accident, and the differences in the velocity of light in different media among the causes of refraction.

Yet another way to doubt the principle, and a particularly prevalent one, stems from a faulty definition of symmetry. It is related to the confusions encouraged by the coordinate-dependent style of formulation mentioned in the last section, and is illustrated by A.F. Chalmers in a 1970 paper. Chalmers writes:

> In a deterministic theory, Curie's Principle will be satisfied for a transformation T if the laws of the theory are invariant under T, for if neither a cause *nor the appropriate laws change under T*, the derivation of the corresponding effect will take an identical form for the transformed and untransformed system and will yield identical expressions for the effect.
> (Chalmers 1970, 133)

Chalmers restricts the application of the principle to transformations which are symmetries of the laws, because he thinks it would otherwise be violated by situations of the following kind: a law which is asymmetric with respect to T takes initial conditions symmetric with respect to T onto final conditions *not* so symmetric. For example, dynamical laws asymmetric under spatial reflection will evolve an experimental set-up which is symmetric under reflection onto one that is not. Suppose, we place a bit of cobalt-60 in a magnetic field created by a ring current in a wire, and two feet away, the mirror image of the (cobalt+ring) set-up. The combined system [(cobalt+ring)$_r$+(cobalt+ring)$_l$] is symmetric with respect to reflection through the plane P separating the right system from the left, but the laws predict that in both subsystems the cobalt-60 atoms will decay significantly more often on the right side. This means that the evolution of the combined system will not be symmetric with respect to reflection through P, for the reflection of the right system should show a prevalence of decay on the left, instead of the right.

Is this a situation in which Curie's Principle is violated? No. A moment of thought should convince one that insofar as the laws predict the lopsided result, the initial experimental set-up cannot be symmetric under reflection. The theoretical description of the (cobalt+ring)$_r$ and (cobalt+ring)$_l$ cannot be the same because the two will evolve differently and the differences in the evolution are predicted by the laws (prevalence of decay *in* the direction of the field for one, and *against* the direction of the field for the other). Chalmers' confusion is the result of his definition of symmetry. Immediately preceding the passage quoted above he writes:

> Any transformation of the co-ordinate system that leaves the mathematical form of a law unchanged is a symmetry (or invariance transformation) for this law. (Chalmers 1970, 144)

He is falling prey here to a mistake that has a long and venerable history. Since it has been clearly discussed and diagnosed by others, I will be short with it.[8] Chalmers' definition identifies the symmetries of a law with the transformations in its covariance group, i.e. the group of those transformations which preserve its truth. I mentioned above that there are two ways of formulating dynamical theories:

(i) in the traditional style which makes use of coordinate systems and with respect to which geometric transformations are expressed as transformations between coordinate systems, and

(ii) in a coordinate independent-style in which geometric transformations take the form of manifold automorphisms, one-one, suitably continuous and differentiable mappings of neighborhoods of M into M.

---

8. See also, J.L. Anderson, Principles of Relativity Physics, New York: Academic Press, (1967), p. 75-83; and M. Friedman, Foundations of Space-time Theories, Princeton, NJ: Princeton University Press (1983), p. 46-62.

Any theory whatsoever can be formulated in the latter style, and—so formulated—is covariant under arbitrary transformations. If we accept Chalmers' identification of the symmetries of a set of laws with its covariance group, we will have to say that any law can be written in a form in which it is symmetric with respect to arbitrary transformations. Moreover, we will have to admit that the symmetries of a given law are not invariant under translation from one formulation into a mathematically equivalent one: formulated in a coordinate-independent manner its symmetries will be the group G of all one-one sufficiently continuous and differentiable transformations, and formulated in coordinate-dependent but mathematically equivalent manner its symmetries will form a proper subgroup of G. Clearly this isn't right. The symmetries of a set of equations are the automorphisms of its set of solutions only when that set is well defined, i.e., when it is the same independent of its relation to any coordinate system, and this is so only when it is given a generally covariant formulation. If Chalmers had been thinking in terms of such a formulation, his mistake would have been clear to him because the left and right subsystems would have been distinguished by the value of some parameter which is not invariant under reflection. This, again, is why I suggested we restrict our attention to the generally covariant formulations of theories, in the hope that we can avoid some of the confusions encouraged by the coordinate-dependent style.

6. THE PRINCIPLE IN INDETERMINISTIC CONTEXTS

I have suggested that we read Curie's Principle as an observation about how the symmetries of some effect are related to those of the conditions which determine it, if such there be. The 'if such there be' expresses a significant restriction, for the deterministic theories which ruled physics in Curie's day have been superseded by quantum mechanics together with its notorious *in*determinism. So the question arises: does Curie's Principle have any application where the laws in question are indeterministic, and in particular,

---

9. Indeterministic laws which pick out a range of possible states without assigning probabilities, are conceivable, so this is a special case. But it is a case which covers all indeterministic laws of which we have actual examples in science, and so it is the important one.

does it have any application in the context of quantum phenomena?

The answer is, in both cases, yes. The difference between deterministic and indeterministic laws is that the former map state-descriptions onto state-descriptions whereas the latter map state-descriptions onto *probability functions which define a distribution over the set of possible state-descriptions*.[9] The Curie-effect of an indeterministic law is just the resulting function, and its asymmetries are defined as follows: T is an asymmetry of a probability function p just in case there is some Bi such that $p(B_i) \neq p(TB_i)$. In the last section, I described experiments in which atoms of cobalt-60 placed in a magnetic field show a prevalence of decay on the left over the right. Experiments like these were first carried out by Wu at Columbia in 1956, and prompted the recognition of the asymmetry of the laws with respect to spatial reflection. The context is an indeterministic one, and the recognition of the asymmetry of the laws involves an application of Curie's Principle: the direction of decay is in each case regarded as undetermined, but the laws are assumed to entail a probability distribution over possible directions of decay which explains the relative frequency of directions in a long run of tests. The left/right asymmetric effect which requires explanation, then, is not the direction of any individual decay event but the probability distribution which favors left over right directions of decay. The significantly greater frequency of left-side decays makes for a left/right asymmetry in the Curie-effect, and hence by Curie's Principle, the Curie-cause A of B cannot be intrinsically symmetric under exchange of left and right.

The principle has another role in indeterministic contexts as well, for it suggests a precise criterion for separating the chancy from the law-governed aspects of a system. Define the *coarse-state* of a system governed by an indeterministic law as the most complete state-description B_c such that $p(B_c)=1$. If we call the state-descriptions over which the distribution is given the *fine-states*, the coarse-state determined by a given probability function is the incomplete state-description which includes all and only those features common to all of the fine-states which get a non-zero probability. Now we can say that *transformations which are symmetries of the coarse state but not symmetries of the fine states will in general be permutations of chancy features of the system*. Consider, for instance, the indeterministic process in which an alpha-particle is emitted from a radioactive nucleus and suppose that the state of the nucleus is such that the particle is no more likely to be emitted in any one direction than another. The coarse state of the nucleus is in this case spherically symmetrical but the fine-states are not, since each of them specifies a particular direction for emission. Those aspects of the fine-states which break the spherical symmetry of the coarse state, i.e., the direction in which the particle is actually emitted, are just its unpredictable aspects.

So let it no longer be said that Curie's Principle doesn't apply in indeterministic contexts; it has all the force it does in deterministic contexts, and more. Not only does it relate the asymmetries of the probability distribution over possible final states which acts as the Curie-cause in indeterministic contexts to those of the initial state, but it provides a precise criterion for separating the chancy from the law-governed features of the final state.

Showing how it applies specifically to quantum mechanics is somewhat complicated because of interpretive disputes about the theory; the best we can do is to divide interpretations into two classes and say how Curie's Principle applies to the interpretations in each class. On interpretations which incorporate a so-called 'collapse postulate', the state of an isolated system evolves deterministically except during 'measurement' when it 'collapses' into a state probabilistically determined by its state before the interaction.[10] So on interpretations of quantum mechanics which incorporate a collapse postulate, the law which determines the probability distribution over the outcomes of measurement interactions for systems given the state of the system before interaction, i.e., Born's rule, is treated as an indeterministic dynamical law. On

---

10. It is a notoriously hard problem for all of the interpretations in this class to characterize measurement interactions precisely, but for the purposes of the discussion, we can assume the problem has a solution.

11. Here is how this characterization applies to some of the more familiar no-collapse interpretations: on the Kochen/Healey/Dieks interpretations, the set of observables for which a composite (measured system + measuring apparatus) system has values, over and above those of which it is in an eigenstate, is determined by the unique biorthogonal decomposition of the system's state (if such there be, i.e., if there is no degeneracy), and the values the system has for those observables is probabilistically determined by applying Born's rule to its Y-function. On the modal interpretation of van Fraassen, the additional observables are determined by the history of the combined system (whether they have just engaged in a measurement interaction and which observable was measured, under some precise physical characterization of the conditions under which these obtain), and - as above - the value those observables have are probabilistically determined by applying Born's rule to its Y-function.

interpretations of quantum mechanics which incorporate no collapse postulate, by contrast, Born's rule is treated as a law of coexistence relating *partial* state descriptions to a probability distribution over fuller state descriptions. The former include only the values of observables of which the system is in an eigenstate, the latter include the values of additional observables.[11] In either case, i.e., whether the Born probability of a value **a** for an observable **A** for a system in a state Y is interpreted as the probability that if we measure **A** it will evolve into an eigenstate of **A** with eigenvalue a or as the probability that the system in fact possesses value a for A, Born's rule is treated as an indeterministic law which maps the state-descriptions onto a probability distribution over state-descriptions. And in either case, the former is the Curie-cause of the latter, and Curie's Principle applies to them just as it does in deterministic contexts.

As to separating the chancy from the law-governed aspects of a quantum-mechanical system; a system's coarse-state includes values for all observables of which it is in an eigenstate and is symmetric under permutations of the values of observables of which it is not. Its fine-states, on the other hand, include values for observables of which it is not in an eigenstate and hence are not symmetric under permutations of those values. Moreover, there are no other transformations which are asymmetries of the fine-states and are not asymmetries of the coarse-state. Applying our criterion for separating the chancy from the law-governed aspects to quantum mechanical systems, then, we get that the values which a system possesses (or comes to possess after a measurement of the right sort) for observables of which it is not in an eigenstate are chancy, and are—moreover—the *only* chancy features of its state. This is just the right result.

7. SPONTANEOUS SYMMETRY BREAKING

Having given an argument to the effect that spontaneous symmetry breaking is impossible, I had better consider the actual phenomena physicists call by that name and show that they do not provide counterexamples. The most striking example of abrupt symmetry change, and also one of the first to be discovered, is Euler's load: if we take a vertical column and load it from the top with a perfectly symmetrical distribution of weight, when the load reaches a critical value, the rod will buckle breaking the symmetry of the initial conditions. Other examples include the change of symmetry that occurs in a liquid uniformly heated from below when the vertical temperature gradient reaches a critical value, an effect demonstrated quite indisputably in experiments performed by Benard—a colleague of Curie in Paris—in 1900. Poincare—also a Paris colleague of Curie's—discovered asymmetric pear-shapes of fast rotating fluid masses in self-gravitating equilibrium... just to mention a couple of instances which were around and available to Curie long before the instances of spontaneous symmetry breaking (usually in the contexts of superconductivity and quantum field theory) which occupy physicists nowadays. In these cases, as with the buckling rod, the symmetry breaks down when a scalar parameter reaches a critical value, without the intervention of visibly asymmetric causes, and the symmetry of the initial state is larger than that of the resulting state. But in light of what preceded, unless these are indeterministic phenomena, Curie's Principle is violated. What is going on?

The clue to the explanation lies in the phrase 'without the intervention of visibly asymmetrical causes'. The case of Euler's load, though striking, is somewhat different than the others. Here, the buckling of the rod in a particular direction appears to be a genuinely indeterministic phenomenon. When the load reaches a particular weight, the differential equations which express the functional dependence of the behavior of the column on the load simply break down. The column buckles, and—so long as the situation is perfectly rotationally symmetric—there is no predicting beforehand the direction in which it will go. The other cases are more interesting, and result from the non-linearity of the dynamics.[12] With respect to actual physical systems, the symmetry of both causes and effects is seldom exact: in part because of the impossibility of completely decoupling a system from its surroundings, in part because of the inevitable presence of experimental error, and in part because of the presence of fluctuations (whether thermal or quantum mechanical). Where the dynamics are governed by non-linear equations, it is *not* true that nearly

---

12. The role of non-linearity explains why all of the early examples come from fluid dynamics. It is an interesting historical fact that although they were being investigated in close proximity to Curie, and close to the time he wrote the symmetry paper, neither he nor his colleagues appear to have considered them in the light of their symmetry properties.

symmetric causes produce nearly symmetric effects or *vice versa*. On the contrary, not only can large asymmetrical causes produce small (even undetectable) effects, but large asymmetric effects may arise from seemingly symmetrical causes. Seemingly symmetrical, but not entirely so; there are always the fluctuations. What happens in every case of so-called spontaneous symmetry breaking in current physics is that when some (usually scalar) parameter reaches a critical value, the symmetric solutions cease to be stable under small external chance perturbations, and so pushing the parameter past the critical value will make the system vulnerable to perturbations that carry it into an asymmetrical state. The symmetrical solution always exists but is not always stable for all values of the parameters that enter the equations of motion. Where it is not, asymmetrical solutions may appear that are stable, thus precipitating a change in the symmetry of the system that appears to be occasioned solely by the change of a scalar parameter. In general, if a system is non-linear and possesses bifurcation points where a set of stable solutions of lower symmetry branch off from the original symmetrical solution and the system is subject to external chance perturbations, a very small chance perturbation may switch the solution to an asymmetrical one.

It should be clear that there is nothing in this which violates Curie's Principle. These are not cases of systems governed by deterministic equations in which all symmetries of causally relevant factors are not preserved in the effect, but rather cases in which the effect of certain causally relevant factors (the external perturbations) appear only under certain conditions, *viz.* when some other parameter reaches a critical value. The situation is commonplace but particularly striking in non-linear contexts where the causes are so small and the effects so large; in such cases, the cause is so apparently symmetrical and the effect so evidently not so, but the asymmetries in the effects are still present in the causes.

8. THE PRINCIPLE AND HIDDEN VARIABLES

No discussion of Curie's Principle would be complete without saying something about its connection with hidden variables. If A is the Curie-cause of B, A is fixed (up to a constant) by the values of the parameters on the left-hand side of the B-determining equations in a generally covariant formulation. Curie's Principle entails that if T is a characteristic asymmetry of B, at least one of the parameters which characterizes A is characteristically asymmetric with respect to T. Call T a **hidden characteristic asymmetry** of A just in case T is an asymmetry of A, but the specifications Ai and TAi are not all distinguishable by unimplemented sight, i.e. in case, for all Ai, we cannot commonly distinguish A_i from TA_i 'just by looking'.[13] And call T an **apparent characteristic asymmetry** just in case it is an asymmetry which is not hidden. It is certainly possible for apparently T-symmetric states to evolve deterministically into apparently T-*a*symmetric states, but Curie's Principle entails that in every such case T is a hidden characteristic asymmetry of the initial state. Hence, if an isolated system in an apparently symmetric state evolves into a state that is evidently T-asymmetric, so long as the evolution is deterministic we can conclude that the symmetric facade of the initial state was a dupe: it concealed all of the characteristic asymmetries revealed in the final state. Molecular biology is replete with particularly impressive instances of apparently T-symmetric conditions giving rise, by evidently deterministic processes, to apparently T-asymmetric effects. Frog zygotes, for example, start out as spherical cells suspended in an homogenous seeming fluid and develop into highly structured organisms; almost every stage in their development introduces asymmetries not apparently present in the preceding stage. By Curie's Principle, if the process is deterministic, the initial state is—despite appearances—at least as asymmetric as the final.[14]

If we combine Curie's Principle with a methodological principle to the effect that the only theoretical reason for postulating a hidden characteristic asymmetry is as the nomological determinant of some apparent asymmetry, we can derive a great deal about

---

13. 'Commonly', here, means something like 'in normal circumstances, with normal vision, with due attention and training, etc.'.
14. We can think of physical evolution, in fact, as precisely a process wherein hidden structure is made apparent. The distinction between hidden and apparent asymmetry is - as I understand it, and in terms more familiar than his own - David Bohm's distinction between explicate and implicate order, and conceiving of evolution in these terms is to think of it as what he refers to as explication of implicate order.
15. If T is not a symmetry of B, then according to Curie's Principle, there is some parameter P which is not invariant under T, and is such that A1—->B1 , A2—->TB1, and A and A' are distinguished by different values of P. If A and A' are not observationally distinguishable, P is a 'hidden variable'. Whenever we postulate hidden variables as the causes of differences in

Curie's Principle 49

how the apparent structure of the physical world relates to the underlying physical structure postulated by our theories. For together they entail that the hidden causes postulated by our theories should collectively conceal as much asymmetry as is revealed in their collective effects and *only* so much asymmetry as is revealed in their collective effects.[15] This narrows the class of theories quite drastically, but doesn't come close to picking out a unique one. A theory will typically regard a single hidden parameter as causally implicated in a wide range of distinct effects while being only partially responsible for any given one. So even if we know that the asymmetries in any effect are present in its Curie-cause and we require that Curie-causes have no hidden asymmetries that are not asymmetries of their effects, there is still room for differences in theory in the way in which the asymmetries get divided up and distributed among the individual parameters. There are better ways of doing the dividing up, of course, but what they are is a question to which neither Curie's Principle nor the above methodological principle, answers.

9. EXEGETICAL SUPPORT

There is some unfinished business; I need to address the question of whether the reading of Curie's Principle that I have given captures Curie's own intentions. Curie writes that "when certain causes produce certain effects, the symmetry elements of the causes must be found in the produced effects". This raises two exegetical questions: (i) what is meant by 'cause' and 'effect'? and (ii) how do we define the symmetries of the causes and effects, so construed? I have read the principle as asserting that the characteris-

---

observable behavior, we assume that there exists a Curie-cause for some family of alternative events and derive the characteristic asymmetries of the postulated cause from those of the effect. It is generally accepted that postulated hidden parameters must be measurable, i.e. must have observable effects beyond the effect it is introduced to explain (e.g., on a measuring apparatus). Without such a constraint, the distinction between deterministic and indeterministically evolving systems becomes empirically empty, for we can postulate hidden differences wherever we observe hidden behavior.

All of this suggests that the specifically theoretical aspect of scientific theorizing (insofar as it consists in choosing between models which embed the same empirical substructures but differ in their higher level theoretical structure) is just an application of Curie's Principle on a cosmic scale, for it is entirely a matter of deriving intrinsic asymmetries in a postulated Curie-cause (the underlying physical structure of the world) from asymmetries in its effect (the appearances), on the assumption that the former determines the latter.

tic symmetries of a Curie-cause are also characteristic symmetries of its effect, i.e., as asserting that if the laws provide a many-one mapping of a family $A = \{A_1, A_2, A_3, ...\}$ of alternative events into a family $B = \{B_1, B_2, B_3, ...\}$, then the transformations which are symmetries of all of the A_i's are also symmetries of all of the B_i's. So far as I can see, no other reading of the principle is compatible with the examples with which Curie illustrates it and the application which he makes of it in his paper.

The paper is quite beautiful; in it Curie does much more than simply state his principle, he uses it to derive the intrinsic symmetries of the electric and magnetic field quantities from the description of certain carefully chosen experimental phenomena. The application he makes of the principle in these arguments leaves little doubt as to how he intends it to be understood. All of the arguments have the same form: a quantity **q** is shown to coincide with a group **G** by application of Curie's Principle to experimental situations in which **q** figures as part of the effect of some phenomenon and the cause of some other. The first is used to establish that the symmetries of **q** include all of the transformations in **G**, the second is used to establish that the symmetries of **q** include no transformations not in **G**. The conclusion is that the group of intrinsic symmetries of **q** is exactly **G**.

The argument deriving the intrinsic symmetry of the electric field at a point, for example, goes like this. Consider the groups of symmetries associated, respectively, with

(A) a cylinder at rest,
(B) a system consisting of two coaxial cylinders rotating in opposite directions,
(C) an arrow, and
(D) a cylinder rotating about its axis.

Assume that mass is a scalar quantity, that linear momentum has the symmetry of (C), and that angular momentum has the symmetry of (D). Now, consider the electric field at the center of a parallel plate condenser made up of two oppositely charged discs. The cause of the field has the symmetry of (C), with the line through the center of the disks as the isotropic axis. From Curie's Principle it follows that the symmetry group of the field at the center of the condenser must include the transformations in (C). Next, consider a point charge placed at a point p in space where there is an electric field. The charge will experience a force which has symmetry (C), and so—again from Curie's Principle—it follows that the symmetry group of the field at p includes *only* transformations in (C). These together entail that the intrinsic symmetry group of the electric field at a point is just (C).

The argument that the intrinsic symmetry of an electric current is also (C) follows the same pattern. Curie considers two situations, one in which a current is the effect of an electric field and another in which a current is the cause of chemical decomposition in electrolysis. Application of Curie's Principle to the former establishes that the symmetry of the current is *at most* (C), application of Curie's Principle to the latter establishes that the symmetry of the current is at *least* (C), and it follows that the symmetry of an electric current is exactly (C). So, also, goes the argument that the magnetic field at a point has the intrinsic symmetry of (D); the relevant experimental situations in this case are those in which a magnetic field is caused by a current in a circular coil and causes electromagnetic induction, respectively.

In all of these arguments, and in every example Curie discusses in other sections of the paper, Curie's Principle is applied to conditions related only as Curie-cause to Curie-effect. Curie is quite explicit that this is what he intends; he writes "whenever a physical phenomenon is expressed as an equation, there is a causal relation between the quantities appearing in both terms." Moreover, it is clear that this is all that is required for the validity of the arguments. There remains the question of whether the symmetries in question are symmetries of particular specifications of Curie-causes and—effects or characteristic symmetries of the families as wholes, i.e. symmetries of *each* specification in the two families, respectively. There are two reasons for thinking that Curie intends the latter. The first is interpretive charity; if we read him as meaning the former, the principle is false. Indeed, quite obviously so; the example of bald and hairy particles given in section 2 is a clear counterexample and it is easy to generate others. The second reason is that in the above arguments, the existence of *any* specification of A that is T-asymmetric is sufficient to establish the T-asymmetry of A, the existence of *any* specification of B that is T-asymmetric is sufficient to establish the T-asymmetry of B, and Curie's Principle is taken to entail that the asymmetries of B (so established), are also asymmetries of A (so established). Curie finds one among the possible specifications A_j of A which is T-asymmetric and one among the possible specifications Bk of B that is T-asymmetric, and concludes that A and B are both T-asymmetric. He does not require that the particular T-asymmetric specifications invoked be related as individual cause and effect, i.e. he does not require that $<A_j, B_k>$ is itself a solution to the laws. Curie's Principle is often read as though it did, i.e. as though it states that the symmetries (idiosyncratic and characteristic alike) of each particular specification of a Curie-cause must also be present in the specifications of its corresponding effect. And indeed, the principle—separated from its context—is ambiguous between such a reading and the reading I have given, but a look at Curie's examples and his own application of the principle in the paper is quite enough to clear up the ambiguity. There can be little serious doubt as to his real intentions.

10. IS THE PRINCIPLE TRIVIAL?

I will take a moment before concluding to fend off the objection that CP—as I've interpreted it—is trivial. There is one sense in which it *is* trivial, namely that it follows from the definitions of the terms in which it is stated. This sort of triviality, however, does not render it insignificant, for it must be granted that there are mathematical theorems which both follow from the definitions of their terms and have a great degree of physical import. A more serious worry is that CP is not only trivial but too *obvious* to be interesting. It doesn't require a long proof or reveal subtle and unexpected connections; one scarcely needs to unpack the definitions to see that it is true. The right way to answer this, I think, is to say that therein lies its beauty. Here is an analogy: any plane figure can be represented in either of two ways, by a line diagram on a page or by the function which generates it. Some truths about the relations between such figures require complicated algebraic proofs but are conspicuously true if we look at the corresponding line diagrams, e.g., the fact that figure A is embeddable in figure B. Other truths about relations between figures are almost impossible to discern from their line drawings but are easy to see with a quick peek at the functions which describe them, e.g., the fact that A results from the composition of functions which generate B and C. One has an elegant way of representing a type of object when the most *important* truths about those objects appear obvious when they are so represented. It is nothing more than a recommendation of conceiving of physical laws as a function from one set into another and attending to the characteristic symmetries of the elements in the two sets, that CP appears obvious when one does. The substance of the principle derives from the fact that the simple relation it expresses between the symmetries of the domain and range of a function can often be used to draw conclusions about the former which cannot be gleaned directly from observation, *from* information about the latter which can. It lies not in seeing that the principle is true, but in recognizing phenomena in the physical world which are related as Curie-cause to—effect, and applying CP to draw conclusions from the *observed* asymmetries of the latter about the often hidden asymmetries of the former.

Some instances of such applications that I have mentioned are these:

(i) the case of the development of frog zygotes described in section 8 in which the apparent determinism of the process, combined with information about the evident geometric asymmetries of the final stage, tells us a great deal about the hidden structure of the apparently spherically symmetric early stages,

(ii) Curie's arguments, described in section 9, deriving the characteristic symmetries of the electromagnetic field quantities, and finally,

(iii) also touched on in section 9, the characteristic asymmetries of the appearances, represented by the empirical substructures of a theory's models, must also be characteristic asymmetries of the underlying physical structure postulated by theories and represented by their higher-level structure. This expresses the only *a priori* constraint that the appearances place on the higher level theoretical structure of our models.

11. CONCLUSION

In the year before Curie's paper was published in the *Journal de Physique*, Sophus Lie published the third and final volume of his *Theorie der Transformationsgruppen* in the preface to which he urged deliberate attention to the symmetries of physical laws. Though Lie and Curie were contemporaries, their emphases were quite different; Lie focused on the symmetries as mathematical properties of the laws, whereas Curie concentrated on the characteristic symmetries of the physical states themselves. In the century separating us from the years in which those seminal works came out, physicists have followed Lie's mathematical emphasis and Curie's basic physical insight has been all but lost.

The insight, as I've suggested, was beautifully simple and quite obvious once one adopted Curie's perspective: a physical law can be identified with a function from one set of physical states into another. If the law is a dynamical law, the states in question are the states of a system at two different times; if it is a law of coexistence, they are partial state descriptions of a system at a single time. Curie's Principle follows just from the notion of such a function; it says that all transformations which are characteristic symmetries of the former are also characteristic symmetries of the latter. More intuitively, it says that transformations which leave the values of all relevant parameters unchanged also leave unchanged their effects. We have seen physical applications of the principle ranging from guiding the postulation of hidden variables in theorizing to providing a criterion for separating out the chancy aspects of a system's evolution. I have not said much about its broader philosophical import, but it should be clear that the way symmetries operate in physics suggests an understanding of the scientific applications of some philosophically important notions. It suggests, in particular, that A is **causally relevant** to B just in case the B-determining laws are not symmetric with respect to arbitrary permutations of the values of A, and that **if (counterfactually)** A_i had occurred, then B_i would have occurred as well; moreover, if A_i had occurred, then no *other* Bi would have. In sum, echoing the first sentence of Curie's paper, I think that there is much interest in introducing into philosophy the symmetry considerations familiar to physicists, and Curie's Principle is a very good place to start.

APPENDIX: GEOMETRIC TRANSFORMATIONS

Let me say a little more than I did in the text about geometric transformations, the way in which the coordinate-dependent style of formulation obscures their physical significance, and how Curie's Principle applies to them. A dynamical theory is typically presented as a state space together with a set of laws; if the laws are deterministic, they pick out a group (or semi-group) of evolution operators $\{U_d\}$ such that if the state at t is Ψ, then the state at t+d is $U_d(\Psi)$. A trajectory through the state space $\Psi(t)$ satisfies the laws just in case for all t and d, $\Psi(t+d)=U_d\Psi(t)$. If T is a transformation defined on the state space,

(i) T is a symmetry of state Ψ *iff* $T\Psi=\Psi$
(ii) T is a symmetry of a set Σ of states *iff* $T\Sigma=_{def}\{T\Psi=\Psi:\Psi$ in $\Sigma\}$
(iii) T is a symmetry of the laws iff for all d, $TU_d=U_dT$, i.e. for all Y, $T(U_d\Psi)=U_d(T\Psi)$. Equivalently, *iff* $U_d=T^{-1}U_dT$ for all d.

Now, clearly if $\Psi=T\Psi$ then $T(U_d\Psi)=TT^{-1}U_dT(\Psi)=U_dT\Psi$, so if T is a symmetry of the initial state and the laws, then it is a symmetry of the final state. What is harder to see, particularly if T is a geometric transformation, but is nevertheless true, is that *if T is not a symmetry of the laws then it is not a symmetry of the set Σ of possible initial states*. Let A be the state at some initial time, and let B be the state at a final time, and suppose T is a geometric transformation which is not a symmetry of the laws. If there exist two physically possible trajectories $<A_i,B_i>$ and $T<A_i,B_i>$ where $B_i \neq TB_i$, it must also be the case that $Ai \neq TAi$. But if there is some Ai such that $A_i \neq TA_i$, then T is not an characteristic symmetry of A, notwithstanding the fact that in the traditional formulation, A_i and TA_i have the same coordinate-dependent description. For those descriptions are given relative to different coordinate systems and there are dynamically relevant differences between the manifolds they represent.

Consider the example I gave in section 4 of an isolated Newtonian system consisting of a ball at rest on a frictionless surface. A_1 is the coordinate-dependent description of the state of the ball at a time t_1 relative to a coordinate system with respect to which the ball is at rest, and B_1 is its state one minute later at t_2; and the ball remains isolated in the interim. $<A_1,B_1>$ is a solution to Newton's laws, and A_1 is the Curie-cause of B_1. The coordinate transformation T, which takes every point (x,y,z,t) onto the corresponding point $(x,y,z+at^2,t)$, carries $<A_1,B_1>$ onto $<TA_1,TB_1>$ has the ball spontaneously accelerating in the z direction with no force acting on it. Newton's laws in their traditional coordinate-dependent representation are not covariant with respect to T, and it is evident in this case that $<TA_1,TB_1>$ does not satisfy them, although $<A_1,B_1>$ does.

What is going on can be described (crudely, but well enough for our purposes) as follows. There are dynamically relevant differences between manifolds represented by Cartesian coordinate-systems moving non-inertially with respect to one another. Hence, in going from the manifold represented by the first coordinate system to the manifold represented by the second, something on which the dynamical behavior of Newtonian systems depends is changed. The relativity to a coordinate system expresses a dependence on parameters which must be explicitly included if one is to formulate laws which hold 'absolutely', i.e., relative to all coordinate systems. This is what is done in the generally covariant, or 'coordinate-independent', representation. Whereas in traditional formulations one states dynamical laws which hold only 'relative to' manifolds with particular intrinsic structures and which are hence covariant only under automorphisms which preserve those structures, in the generally covariant formulations one states laws which explicitly relate dynamical behavior to the relevant intrinsic structures. The new laws hold 'relative to' all manifolds and are consequently covariant with respect to all transformations between them. This is why it is true only of the generally covariant formulation of a theory that if T is a geometric asymmetry of its laws, there is a parameter in the B-determining equations which is not invariant under T. In the same way, if we have laws which hold only 'relative to' systems of one or another mass, in order to formulate laws which hold for systems of any mass, we have to relate the behavior of a system to its mass and a new parameter representing mass (i.e., not invariant under changes in mass) will appear in the equations.

Essay 3
Reflection in Space

1. THE INTERDEFINABILITY OF SYMMETRY, SIMILARITY, AND INTRINSIC PROPERTIES

If we have a set of properties we can talk about the transformations under which they are invariant. Since the set will in general include the Identity, and be closed under inverse and composition, it will have a group structure. In the first essay, we looked at some simple geometric figures in Euclidean space and the corresponding groups of transformations which map each onto itself. This was a loose way of speaking, actually, a transformation h associates each of the points in a space with another; if h is continuous and we consider a set of points in the original space which forms a figure F, then the set of points associated with these by h will form a figure F'. F' will not in general be identical to F but may differ in some respects while remaining the same in others, for example, if F is a square and we consider the translation which takes every point in the containing space onto a point at some fixed distance d in a given direction, F' will be a square congruent to and at a distance d from F. Under the action of a more complicated transformation, F' may be not a square but a rectangle whose length is, say, twice its height. The symmetries of a figure map it onto one that is perfectly similar to it, and the symmetries of a space map any pair of perfectly similar figures contained in it onto another pair, also perfectly similar. We can think of transformations in either of two ways, as inducing a change in a single object, or as associating one object with another. The two ways of thinking are equivalent, but it is well to keep them separate in our own minds; in the former case, the question of whether a particular transformation is a symmetry is the question of whether the transformation doesn't affect intrinsic features of the object; in the latter case, it is the question of whether the associated object is completely similar to the original.

This interdefinability of similarity and symmetry is quite general: a symmetry is a transformation which takes F onto a perfectly similar object F', and F' is similar to F *iff* it can be obtained from F by a symmetry.[1] So if we know the general conditions under which two objects are similar we can determine their symmetries and if we know the symmetries of an object we can determine the conditions under which any other figure is similar to it.[2] The notions of (perfect) similarity and intrinsic properties are also interdefinable: an object is (perfectly) similar to another *iff* it shares all of its intrinsic properties, and a property is an intrinsic property of a figure F *iff* it is shared by all figures perfectly similar to F.[3] This means, of course, that an object obtainable from F by a symmetry shares all of F's intrinsic properties, and no symmetry of F changes F intrinsically. The circle of interdefinability here is small, but the hope of breaking it by giving an independent definition of any one of its members seems bleak. The three questions go hand in hand:

(i) when are two objects are similar?
(ii) under what conditions is a transformation which takes one into the other a symmetry? and
(iii) are the differences between them differences in intrinsic properties?

---

1. I'll suppress the 'perfect' from here on in, but understand it as tacitly included unless otherwise indicated.
2. In general, take any property and consider the set of transformations which preserve that property. These will be the set of transformations which take all and only objects which possess that property onto objects which possess that property, i.e. the set of automorphisms of the set of objects which possess the property in question.
3. We might require in addition that there be objects which don't have the property so that trivial properties (properties shared by all objects, e.g.. the property of being either white or not white) don't turn out to be intrinsic.

2. ENANTIOMORPHS

A context in which the questions come together in a particularly confusing way is in discussion of the reflection in space. Consider my hands, ignoring the incidental differences between them and supposing that each is a perfect mirror image of the other. In almost all respects they are the same: they have the same color, the same parts, their parts have the same lengths and are separated by angles of the same magnitude, all marks on one have a corresponding mark on the other, and so on. In short, aside from the different orientations, they are perfect duplicates. Any sentence we can construct not containing chiral terms ('left', 'right' or terms interdefinable with them) is either true of both or true of neither, but they are nevertheless easily distinguishable. Compare the following cases: (i) you show a friend a pair of dimes which are perfect duplicates of one another, introducing one as 'Joe' and the other as 'Bob', then put them behind your back and shuffle them around. Your friend surely will not be able to tell you which is which. (ii) Do the same thing with a miniature plastic replica of my hands, and no matter how well you shuffle there will be no problem about picking out the left one; it will be immediately recognizable as such. What is it that distinguishes it from the right? Do the two have different intrinsic properties? Not obviously, for it is not obvious that a single hand in an otherwise empty universe is either right or left. There is no measurable geometrical feature which would decide the case, and we are not allowed to import another object whose handedness has been established to decide the case by comparison. Does a hand, considered in isolation from relationships with all other geometrical structures, have some property— some intrinsic feature—in virtue of which it can be said to be either left or right?

There is a temptation, which Kant has been accused of succumbing to, to conclude that it does by reasoning as follows: if a human body, complete but for the absence of hands, were to suddenly materialize near the hand it will only one of the wrists (so that when the palm is against the chest, the thumb points upwards). If it fits the left wrist, it is a left hand and if it fits the right, it is right. But whichever it proves to be, it must have been *before* the body appeared since the mere addition of the body surely didn't produce any change in the hand, hence there must be some basis for calling the hand 'left' or 'right' even when it is alone in the universe. The quick response is that if we don't know which side of the body is left or right, showing that a hand fits on only one of the wrists doesn't go anywhere towards revealing that it is either. The deeper response is that *no* facts about *relations* (counterfactual or actual) of sameness and difference of handedness between objects can yield an answer to the question of whether there are intrinsic differences between the *individual* objects so related. Just as it makes perfect sense to hold that if you introduce a body into the one-handed universe, there will be a fact about the distance between the two without admitting that there is anything in the one-handed universe which determines what that distance will be, i.e., without thinking that there is anything in the *one*-handed universe for 'distance from a body, were one to be introduced' to latch onto (that's just what it means to deny that distance is an internal relation), we can allow that if a wrist were introduced into a one-handed universe, there would be a definite fact about whether the hand will fit it, without admitting that there is anything in the *one*-handed universe for 'orientation with respect to a wrist, were it to be introduced' to latch onto. We can allow, that is, that there is a definite fact about whether a given hand *would* fit a particular wrist, were the two paired, without supposing any intrinsic feature of either which determines the fact.[4] To say that there is a definite fact about whether a given hand *would* fit a particular wrist is to recognize relations of same-handedness. Given such relations, together with an independent way of picking out some asymmetric object O, we can get facts about leftness and rightness on the cheap by defining 'left-handed' as 'being same-handed as O'. This is to construe leftness and rightness implicitly as relations to O. Martin Gardner (in his book *The Ambidextrous Universe*) mentions an episode in John Hart's comic strip B.C. from newspapers of July 26, 1963 that makes the point nicely. One of his cavemen has just invented the drum; he strikes a log with a stick held in one hand and says 'That's a left flam', then hits the log

---

4. "Makes perfect sense" but may not be right. The point here is only that we can accommodate facts about same and different handedness between objects without admitting intrinsic properties of leftness and rightness by construing the former as an external relation. The example of (absolute) spatial separation is just a convenient and familiar relation that is naturally so construed. I don't mean to suggest that space - whether relational or substantival - need be thought as imbued with a definite metric.

with a stick in the other hand and says 'That's a right flam'. Another of the cavemen, who has been watching, asks 'How do you know?' The first, pointing to the back of one of his hands replies 'I have a mole on my left hand'. The thing is funny because the spectator took himself to be asking about a distinguishing intrinsic difference between the right and left flams, and the drummer answered by adverting to the differing relations they bear to his moled hand.

Some mathematical detail: mirror image objects like the hands above, are carried into one another by reflection (more generally, by an improper rotation) are called *counterparts* if they share all of their metrical properties (all of their parts have the same length and magnitude of angles between them), and *incongruous counterparts* if they cannot be brought into congruence (made to occupy the same region of space) by a continuous rigid motion (a combination of rotations and reflections). Since two objects can share all metrical properties and yet be mirror images of one another, leftness and rightness evidently don't supervene on metrical properties, and since such objects can not always be brought into congruence by a continuous rigid motion reflection is not the same as any combination of rotations and translations. This means that supposing—as we do—that metrical properties are intrinsic to the objects which possess them doesn't settle the question of whether leftness and rightness are, and it means that supposing—as we do—that translations and rotations are symmetries of physical space doesn't settle the question of whether reflection is. These, apparently, have to be answered separately.

They have a long history in discussions of the metaphysics of space beginning with Kant's famous 1768 paper, "Concerning the Ultimate foundations of the Differentiation of Regions in Space"[5], in which he argues from the existence of incongruous counterparts to substantivalism about space.[6] The same facts crop up again in the *Prolegomena* and in the *Critique* where they figure somewhat differently (Kant had, by that time, abandoned the belief in absolute

---

5. G.B. Kerferd and D.E. Walford, trans. *Kant: selected Pre-Critical Writings* (Manchester: University Press, 1968).

6. Substantivalism, of course, is the view that space exists as an entity over and above the material objects it contains and their spatial relations to one another. It is traditionally opposed to relationalism, the view that space consists in a network of relations between (actual and possible) material bodies.

space). The earlier paper, because of the headiness of the conclusion and the peculiarity of the argument, has been steeped in controversy since its publication. Recently Johnathan Bennett, Peter Remnant and John Earman have launched attacks on the argument which were regarded by many as decisive.[7] Graham Nerlich has tried to defend it in an ingenious way, and his attempt provoked a spirited response from Sklar.[8] The historical literature on the relations between the 1768 paper and Kant's later writings is daunting, and the relation between questions about the intrinsic similarity of incongruous counterparts and the set of debates that go by the name 'relationalism vs. substantivalism' is complicated and not as straightforward as some have supposed. I will skirt these issues, weaving my way selectively through some of the more important recent arguments with an eye towards answering the question of whether the claim that the lone hand in the one-handed universe is neither left nor right, as Kant insisted, "*contradicts the most obvious experience*". I will suggest that some beg the question while others introduce red herrings, and will try to reconstruct an argument which does neither. Finally, I will turn to some rather less obvious—indeed rather recondite, only recently discovered, and entirely unexpected— facts that stand against the claim, *viz.* the apparent violation of parity in experiments concerning the weak nuclear interaction.

3. KANT, REMNANT, EARMAN, AND NERLICH

Kant's 1768 argument runs something like this: since the internal relationships of parts of left—and right—handed objects are the same, nothing about the intrinsic nature of these objects can serve to distinguish them. Nevertheless, a universe consisting solely of a right hand and one consisting solely of a left hand are clearly different, so the two hands must be distinguished by their respective relationships to a third thing. In a one-handed universe, however,

---

7. J. Bennett, "The Difference Between Right and Left", *American Philosophical Quarterly*, VII, 3, (July 1970): 175-191. P. Remnant, "Incongruent Counterparts and Absolute Space", *Mind*, New Series 72 (1973): 393-399. J. Earman, "Kant, Incongruous Counterparts, and the Nature of Space and Space-Time", *Ratio* 13 (1971): 1-18..
8. G. Nerlich, "Hands, Knees, and Absolute Space", *J.Phil.* 79 (1973): 337-351. L. Sklar, "Incongruous Counterparts, Intrinsic Features, and the Substantivality of Space", *J. Phil.* 71 (1974): 277-290.

there is nothing for the hands to bear such a relationship to except space itself. Thus the existence of incongruous counterparts is testimony to the existence of a single thing present in both universes to which the left and the right hands bear differing relations, and that thing is space itself. The structure of the argument is this:

(i) any hand must be either left or right.
(ii) leftness and rightness are not intrinsic properties of hands.
(iii) leftness and rightness are not internal relational properties among the parts of hands.
(iv) leftness and rightness are not external relations of a hand to parts of space.
(v) a hand is left or right (has its leftness or rightness) in virtue of its relation to space in respect of some property that space possesses as a whole; to have made a left hand rather than a right, Kant says, would have required "a different action of the creative cause", relating the hand differently "to space in general as a unity of which each extension must be regarded as a part."

(i) is simply asserted: Kant seems to regard it as self evident. He does mention that it cannot be indeterminate as to which wrist of a human body a hand would fit correctly, but this just asserts a counterfactual relation to another object and provides no argument for (i). Remnant interprets Kant as proposing method of deciding whether a given hand is right or left, and he quite rightly points out that unless we have already settled which wrist of the body in question is right and which is left, the fact that the hand will fit on one but not both, can do no such thing. To settle which wrist is right, however, is just to settle handedness for a different sort of object, so the invocation of the body does no more than push the question one step back. The point is perfectly general: no facts about *relations* of handedness *between* objects can answer the question of whether the objects in question differ intrinsically, for that is the question of whether similarity or difference of orientation is an internal relation, and nothing that has been said entails that it is. So we can take Kant as asserting (i) as self evident. (ii) is entailed by the fact that all intrinsic properties are preserved under reflection, as are metrical relations between the hand's parts. Since Kant holds that the only internal relational properties of a hand are metrical, this yields (iii). The argument for (iv) is interesting; if the handedness of an object were an external relation it bore to some part of space, then motion through space—which just is the changing of an object's external relations to the parts of space—ought to change its handedness. But—he says—an object retains its handedness however it is moved, so handedness cannot be an external relation to parts of space, and the only remaining option, he says, is that handedness is a relation to space 'taken in general as a unity'. It is hard to make out precisely what is meant, one might have thought Kant is suggesting the handedness is an internal relation between an object and space as a whole, i.e., a relation which supervenes on the intrinsic natures of the two, but since Kant seems to have granted that left and right hands are intrinsically alike, he cannot say that they bear different *internal* relations to a *single* thing.

Here is the problem; the argument looks like it goes like this; all spatial relations for hands are external or internal. The only internal relations are metrical relations, and these are invariant under reflection and hence do not fix handedness. External relations to parts of space consist in position and orientation with respect to points, lines, etc. outside the hand, but these are also invariant under reflection (and in any case changes in these relations occur only if the hand moves through the space and movement through the space doesn't alter handedness). Hence handedness must be a relation to space itself as a whole. In Kant's words

> the positions of the parts of space in relation to one another presuppose the region towards which they are ordered in such relation; and this region, in ultimate analysis, consists not in the relation of one thing in space to another (which is properly the concept of position) but in *the relation of the system of these positions to the absolute world-space*. In anything extended the position of parts relatively to one another can be adequately determined from consideration of the thing itself; but the region towards which this ordering of the parts is directed involves reference to the space outside the thing; not, indeed, to points in this wider space—for this would be nothing else but the position of the parts of the thing in outer relation—but to universal space as a unity of which every extension must be regarded as a part. (Kant, 1768, p. 27, my emphasis)

The problem with the argument is that it is not enough to invoke 'space itself' to explain the difference in handedness, Kant has to name the facts about space and the relations the hands bear to it that distinguish them. Doing so, however, is just the same problem as distinguishing the hands themselves; for a right and a left handed glove each coincide with regions of space and their relations to space can serve to distinguish them only if the regions in question can be distinguished. If we are going to grant that the regions of space occupied by the hands are distinguished intrinsically by their handedness, we might just as well have granted that the hands themselves are so distinguished, for it is no more plausible to suggest that differently handed regions of space differ intrinsically than it is to say differently handed hands do so. But then the existence of incongruous counterparts clearly can provide no argument for absolute space.

That was, essentially, Earman's attack on Kant's argument, his own position is that the hands differ intrinsically. Kant, he says, admitted size of parts and magnitude of angles as intrinsic features of objects, noticed that these don't distinguish differently handed objects, and concluded that handedness must be a relation of some sort. Since he insisted that even a hand alone in the universe would be right or left, there was nothing for it to be a relation *to* but space itself. Rather than adhering dogmatically to the idea that size of parts and magnitude of angles exhaust the intrinsic properties of the hands, he should have concluded that the orientation of their parts was an intrinsic feature in its own right. Thus Earman. There is, however, an interesting complication that forces a retrenchment; in a globally non-orientable space or in a space embedded in one of higher dimensionality, any two counterparts can be brought into congruence by a continuous rigid motion, even though they may not be bringable into congruence by a *local* continuous rigid motion which keeps them confined to the lower dimensional subspace or doesn't take them on a global trajectories. Earman has to admit, if he is to hold onto his position, that in such spaces, non-local motions of the right sort impart changes in the *intrinsic natures* of things.

It is just this kind of consideration which Nerlich exploits in Kant's defense. There is, he thinks, something right in the claim that what distinguishes right and left hands has something to do with space 'considered as a unity' that is borne out modern geometrical account of incongruous counterparts ('enantiomorphs') bears this out. He writes:

> Which of these ... determinate characters [being enantiomorphic (a member of a possible pair of incongruous counterparts), or homomorphic (bringable into coincidence with any congruent object)] a hand bears depends, still, on the nature of the space it inhabits, not on other objects. The nature of this space, whether it is orientable, how many dimensions it has, is absolute and primitive. (p. 345)

> space [is] a definite topological entity [and] can only be a primitive and absolute entity; its nature bestows a character of homomorphism, leftness or whatever it might be, on suitable objects. My conviction of the profundity of Kant's argument rests on my being quite unable to see what the relativist can urge against this, except further relativist dogma (p. 350)

I'm not sure what some of this means (e.g., what it is for space to be 'absolute and primitive'), but the part of the argument which is interesting for our purposes is the claim that handedness cannot be construed as an intrinsic property *of hands*. The argument appears to be this: given a pair of two-dimensional counterparts in the same plane, it may be possible to bring them into congruence by a continuous rigid motion that takes at least one of them out of the plane, but not by any continuous rigid motion that keeps both confined to the plane. Similarly for objects in our apparently three-dimensional space. If our space really is three-dimensional and if it is orientable, then the objects may be such that no continuous rigid motion can bring them into congruence. Yet *had it been* four-dimensional or non-orientable, there would have been a continuous rigid motion which would have brought them into congruence, and in that case the objects would not have been incongruous counterparts. But these sorts of global facts about a space surely cannot be relevant to the intrinsic properties of localized objects like hands, for intrinsic properties are precisely those which an object has in virtue of the way it is *irrespective of all other features of the universe*. If being right-handed and left-handed are intrinsic features of hands, then whether a pair of hands are incongruous counterparts supervenes only on the intrinsic properties of the two. And it follows that whether or not a pair of hands are incongruous counterparts ought not to depend on features of the universe like the global topology of the space in which they are contained. But, Nerlich points out, they do, and the upshot is that being incongruous counterparts cannot be an intrinsic property of the pair, and hence leftness or rightness cannot be intrinsic properties of either individually. We can be somewhat more explicit than Nerlich; the intrinsic properties of an object O are those which are invariant under transformations of the intrinsic properties of distinct objects, as well as the external relations between such objects and O. Whether a hand is a possible member of a pair of incongruous counterparts, and hence whether it is left or right, is not invariant under changes in the topology of the space in which it is contained, hence leftness and rightness are not intrinsic properties of hands. Nerlich doesn't, but should, mention that the argument requires a premise to the effect that the topology of the containing space is either an intrinsic property of an individual distinct from the pair, or that it is an external relation between such existences. If one is a substantivalist about space, then space is an entity distinct from the bodies that occupy it, and its topology is one of its intrinsic properties. If one is a relationalist about space, the topological features of space will turn out to be global facts about the set of external relations between (actual or possible, depending on the species of relationalism) physical objects. In either case, the intrinsic properties of the hands will have to be invariant under changes in the topology of the containing space; either way, Nerlich's premise appears to be okay.

The argument is not, however, successful. Although Nerlich is right that any two three-dimensional hands which are counterparts and which are in the three-space will be bringable into congruence by a continuous rigid motion that takes at least one of them out of the space into other parts of the embedding four-space, there will still be pairs of possible incongruous counterpart hands in the three-space in the sense that the members of the pair are counterparts and no continuous rigid motion that keeps both in the three-space will bring them into congruence. Similarly, even if a space is globally non-orientable there may be a subspace S such that we can divide all the counterpart objects in S into classes so that an object in one class cannot be brought into coincidence counterparts in the other class by a continuous rigid motion *in S*. So even in globally non-orientable spaces, the notion of orientability, or handedness, with respect to a subspace S is perfectly well-defined. Moreover, one can obtain the mirror image of an object in S by reflection, and *we can still ask whether the mirrored object is always intrinsically like the original*. What depends on the topological features of the space that Nerlich is pointing to is the composition of the class of continuous rigid motions and, in particular, whether it includes transformations equivalent to reflection in the subspace in which the two hands are located.[9] If we want to say that the mirror image of handed objects are dissimilar, then we will simply say that Nerlich has only shown that in such spaces the continuous rigid motions are not all symmetries. Nothing he has said rules this out.

To repeat, Nerlich's reasoning went something like this: start by asking whether two handed objects which are mirror images of one another share all intrinsic geometric properties. Call such pairs 'incongruous counterparts', and define them geometrically as counterparts that cannot be brought into congruence by a continuous rigid motion in the space in which they are contained. Notice, now, that if the space in question is globally non-orientable or embedded in a higher dimensional space, the class of continuous rigid motions contains transformations equivalent to reflections, and hence that any objects which are mirror images of one another can be brought into congruence by a continuous rigid motion. Since no intrinsic features of bounded figures should depend on global features of the space, whether a pair are incongruous counterparts of one another cannot be determined by intrinsic features of the pair. This is all true enough, but, again, we cannot conclude therefrom that rightness and leftness are not intrinsic features of handed objects unless the fact that one is left and one is right makes the two incongruous counterparts. But it does not; this is just the upshot of the considerations Nerlich is pointing to. In some spaces, even right and left hands can be brought into congruence by a continuous rigid motion, because in such spaces the continuous rigid motions include transformations equivalent to reflections. "Continuous rigid motion" is a description which picks out different transformations in spaces with different topologies; since enantiomorphs are objects which can't be brought into congruence with their mirror-images by a continuous rigid motion, which objects are enantiomorphs will also depend on the topology of the containing space.

So the question remains; can left and right hands be intrinsically exactly alike? Here are the geometrical facts we have discussed so far:

---

9. I have assumed that the space has constant curvature, but at no cost in philosophical generality.

(i) a pair of congruous counterparts is intrinsically different from a pair of incongruous counterparts (a matched pair of left hands can be distinguished from a left/right pair)
(ii) leftness or rightness does not supervene on metrical properties (hands which share all their metrical properties can differ with respect to handedness)
(iii) reflection is not in general equivalent to a combination of rotations and translations (in orientable spaces, no set of translations and rotations will bring an object into coincidence with its mirror image)
(iv) in some spaces, however, i.e. in non-orientable spaces, a reflection is equivalent to a combination of rotations and translations. And reflection in a space of n dimensions is always equivalent to some rotation in n+1 dimensional space.

Here is what these facts show:
In a left/right asymmetric universe it suffices to distinguish left and right oriented objects (including hands) by picking out one side by an identifying description, dubbing it 'left', and regarding all similarly oriented objects as left, like letting 'left' mean 'same-handed as my moled-hand' in the B.C. cartoon. Kant's strategy of distinguishing left and right by relations to a third object fails not because they cannot be so distinguished, but because to play that role, the third object must itself be left/right asymmetric, we must be able to identify one side by some independent non-chiral property. 'Space itself considered as a unity', as Kant conceived it, possessed no such asymmetry. Neither will relations between material bodies (actual or counterfactual), like the fact that a right hand will fit properly on one but not both wrists of a human body or the fact that a glove will fit only one of a pair of opposite hands, do the trick of distinguishing left from right intrinsically. For this establishes only the existence of relations of different handedness, and—as we've seen—these don't necessarily make for intrinsic differences between the related objects. Nor as we saw in the last section—does the possibility of non-orientable spaces or spaces of higher dimensionality which contain subspaces isomorphic to the region of actual space which contains a given hand, bear directly on the question of what distinguishes left and right. So far as all these considerations go, the two options remain:

(i) to distinguish the two intrinsically by recognizing handedness as a primitive intrinsic geometrical property which, as it happens, fails to supervene on length of parts and magnitude of angles between them, or

(ii) to regard the two as intrinsically similar but related by an external relation ('same or different handedness') which fails to supervene on relations of spatiotemporal separation.[10]

4. A BETTER ARGUMENT

Let's see if we can come up with an argument that fares a little better than the ones we have discussed. It is true that handedness does not supervene on metrical properties and relations, and that reflection preserves these, but this only impugns the intrinsicality of handedness if we already have a reason for thinking only metrical properties and relations are intrinsic. Similarly, we can't get mileage out of the fact that reflection is not in general equivalent to any combination of rotations and translations in space without independent reason for thinking that these latter are symmetries.

I think we *do*, however, without speculating for the moment on its source, have an independent intuition that mere physical motions do not disturb the intrinsic properties of objects, and I think we can come up with an argument somewhat like Nerlich's which explicitly appeals to this intuition by reasoning as follows:

---

10. An internal relation is one that supervenes on the intrinsic natures of its relata [e.g., being larger than, or a different color than]; if X_1 and Y_1 stand in the relation but X_2 and Y_2 do not, then there must be a difference in intrinsic nature either between the X's or else between the Y's. If X_1 and X_2 are perfectly similar, and so are Y_1 and Y_2, then the pairs $<X_1,Y_1>$ and $<X_2,Y_2>$ stand in exactly the same internal relations. External relations, like relations of spatiotemporal distance, do not supervene on the natures of the relata. If are X_1 and X_2 perfectly similar, and so are Y_1 and Y_2, it may yet happen that the pairs $<X_1,Y_1>$ and $<X_2,Y_2>$ stand in different relations of distance. Although distance fails to supervene on the intrinsic natures of the relata taken separately, it does supervene on the intrinsic nature of the complex system consisting of the pair.

(i) if certain conditions obtain, there are *physical motions* which take an object onto its mirror image, and under such conditions reflections are symmetries,
(ii) whether or not the relevant conditions obtain has nothing to do with the local conditions which determine the intrinsic properties of the objects,
(iii) therefore, even in situations where those conditions don't obtain, as (we suppose) in the actual world, reflections are symmetries.

The first premise asserts a *counterfactual*: If certain (to the best of our knowledge) counterfactual circumstances obtain, then reflections are symmetries.[11] The second argues that the difference between worlds in which the antecedent obtains and worlds in which it doesn't is not a difference in the intrinsic geometric properties of physical objects, and hence we can detach the consequent, obtaining (iii). We can argue for (ii) as Nerlich does, by pointing to the fact that the changes in question are changes in the local topological properties of the spaces of the two worlds, and assume that intrinsic geometric properties are local.[12] So the idea is that mere motions do not affect the intrinsic properties of physical objects, continuous rigid motions are the mathematical corollaries of physical motions; our space is, as a matter of fact, orientable, but there is some world w which is locally just like ours in which space is either non-orientable or a subspace of a higher-dimensional space. In w, reflections are (equivalent to) continuous rigid motions and we can take objects into their mirror images merely by moving them around in the space. Since (mere) motions don't affect the intrinsic nature of physical objects, in w objects and their mirror images are duplicates. But the intrinsic properties of objects like hands and gloves are the same in our world as in w, hence in our world just as much as in w, hands are duplicates of

---

11. Actually, our world may be such a world. The evidence suggests otherwise, but the evidence for the global structure of space-time is meager.
12. A little more technically, we can describe a hand as a mass of organic matter extended in a certain metrical three-space where the three-space is bounded by extremal elements that make up surfaces which define the shape of the mass is extended in by limiting it. Now we can talk about the *spaces in which a duplicate of this hand might exist* as the spaces which can embed subspaces isomorphic to this handspace, and we notice that such spaces include ones that differ topologically with respect to orientability and dimensionality.

their mirror images. In our world, as in w, there is no intrinsic difference between right and left hands.

The reasoning can be made especially persuasive if we start by considering objects in two dimensions where there is an *actual* motion (*viz.*, rotation through the third dimension) corresponding to reflection:

> Imagine counterpart angled shapes ['knees', illustrated below] cut out of paper ... They lie on a large table. As I look down on two of them, the thick bar of each knee points away from me, but the thin bar of one points to my left, the thin bar of the other to my right. Though they are counterparts, it is obvious that no rigid motion of the first which confines it to the table's surface, can map it into its counterpart, the second knee...Their being enantiomorphic clearly depends on confining the rigid motions to the space of the table top, or the Euclidean plane. Lift a knee up and turn it over, through a rigid motion in three-space, and it returns to the table as a congruous counterpart of its mate. That the

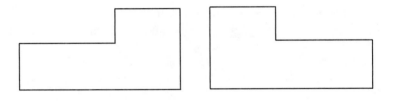

> knees were incongruous depended on how they were put on the table or how they were entered *in* the space to which we confined their rigid motions. (Nerlich, p. 344)

It just seems obvious that the knees are not geometrically altered by picking them up off the table and replacing them face down. To a Flatlander living in the surface who saw the object disappear to be replaced by it's mirror image, it would look as though the object underwent some spontaneous change, but this is just an artifact of the Flatlander's limited perspective. We, who can see that the object merely underwent a physical motion, are quite sure that it is intrinsically unaltered. In two dimensions, figures are mirrored by a line; in three dimensions, by a plane; in four dimensions, by a solid; and in general in a space of n dimensions objects can be reflected through a 'surface' of n-1 dimensions, and in every space of n dimensions an asymmetric figure can be made to coincide with its reflection by rotating it through a space of n+1 dimensions. So if our three-space were embedded in a four-dimensional space, beings occupying the space could simply 'pick objects up' and rotate them through the fourth dimension to take them into their mirror images in the three dimensional space we occupy. It would be obvious to such beings that reflections in our space do not alter objects, and we ought to conclude likewise. That, at least, is what is suggested by this line of thought.

The proponent of intrinsic properties of left and right is precisely denying that reflection is a symmetry and when it is pointed out that reflection in some space can be mimicked by one or another sort of transformation, she will simply deny that the latter is a symmetry either. To the extent that we are looking for an argument that doesn't take as a premise the denial of the view that we mean to be arguing against, we must have independent reason for thinking that the mimicking transformation is a symmetry, i.e. that it is the kind of transformation which doesn't alter the intrinsic nature of the transformed object. Furthermore, the argument will be only as strong as the independent intuition. Here, we are showing that reflection in n dimensions is equivalent to a possible physical motion in n+1 dimensions, and the independent intuition is that physical motions are symmetries. The force of the argument depends entirely on the strength of this intuition. It is clearly not knockdown since there is no *conceptual* bar to thinking that physical motions alter the intrinsic properties of objects. Unfortunately, I don't even think it is strong, for the intuition is quite fragile. One illustration of this is that many people, when they first learn Special Relativity, rather than thinking in terms of Minkowski space-time, find it more intuitive to think of Lorentz contraction as a physical process: objects shrink in proportion as their velocities approach the speed of light.

There is a related argument in which the transformation mimicking reflection is not a rotation through a higher dimensional space of the figures in question but a transformation in the position of a viewer in the higher dimensional space which alters his relationship to the figures in the subspace.

> Imagine a Flatland on a vertical sheet of glass standing in the center of a room. It is, say, a left-handed world when you view it from one side of the glass. Walk around the glass, you see it as a right-handed world ... Flatland itself does not change in any way when you view it from another side. The only change is in the spatial relation, in 3-space, of Flatland and you. In precisely the same way, an inhabitant of 4-space could view one of our kitchen corkscrews from one side and see a right-handed helix, then change his position and see the same corkscrew from the other side as a left-handed helix. (Gardner, "The Fourth Dimension", in *The Ambidextrous Universe*, p. 69)

The form of the argument is the same, but it is a little harder to know what to make of it. There are science fiction stories of people who get somehow 'turned inside out' so that everything appears to them suddenly as though it were seen in a mirror.[13] One will say these are stories about a sudden change in the world or a sudden change in the person's *relation* to the world depending on whether one takes left and right to be intrinsic. To the extent that there is an *independent* intuition in the argument above, it stems from the conviction that the transformation—being a mere change in the position of the *viewer*—alters nothing but the extrinsic spatial relation between she and the figures embedded in the subspace. This is a better argument than the last, and in fact the best that I can see, but it's persuasive force depends, again, on how much weight one gives to the intuition. One of the interesting things about this issue is that we have a question 'are objects duplicates of their mirror images, or do they differ in some (intrinsic) way?', divided intuitions about the answer, a couple of equivalent questions about which we also have divided intuitions, and apparently no independent way to decide the issue. The back and forth inevitably takes the form of intuition mongering.

For all that we've said so far, the question still seems to me up for grabs, but it turns out that there are some rather surprising and unexpected physical considerations that may decide the case in favor of recognizing leftness and rightness as intrinsic geometric properties.

---

13. For example: "When Mrs. Smith started to push open the glass door at the entrance to the bank, she was puzzled to see the world **TUO** printed on the door in large black letters. What does the word mean?" (Gardner, ibid. p. 69)

5. SYMMETRY AS A CONSTRAINT ON MEANING

Before I go on to these, however, let me shift gears and take up a common way of presenting the issue here that is suggested by some of Kant's remarks about incongruous counterparts in the *Inaugural Dissertation*[14], and which Bennett, Gardner, Feynman, and others employ. The idea is to set it up as a problem of how to convey the meaning of the terms 'left' and 'right' to a population who doesn't already possess any chiral terms (e.g., 'clockwise', 'right-handed screw', etc.), and with whom we share no experience (so we cannot define 'right' by ostension). "Is there any way", Gardner asks,

> to communicate the meaning of 'left' [to the inhabitants of some Planet X in a distant galaxy] by a language transmitted in the form of pulsating signals? By the terms of the problem we may say anything we please to our listeners, ask them to perform any experiment whatever, with one proviso: There is to be no asymmetric object or structure that we and they can observe in common. (*ibid*, p. 70)

Bennett presents a closely related problem: imagine that, instead of not knowing what we mean by 'left' and 'right', the inhabitants of Planet X ('alphans') have got the two switched around. Is there anything we could tell them to make them discover their error? Bennett and Gardner each go through a variety of strategies that don't work, and then argue that—except by appeal to the recondite physical phenomena which are the subject matter of the next section—there is not. All of this is intended to show that the difference between left and right is either nonexistent or mysterious. In reality, it does no more than illustrate in a vivid manner that differences in chiral properties do not supervene on, and hence are not stateable in terms of, differences in metrical properties, but this is just a consequence of the fact that all such properties are preserved by reflection. The air of mystery comes from the tacit assumption that such geometric differences as there

---

14. "We cannot by any sharpness of intellect *describe discursively*, that is by intellectual marks, the distinction in a given space between things which lie towards one quarter and things which are turned towards the opposite quarter. Thus if we take solids completely equal and similar but incongruent, such as the right and left hands (so far as they are conceived only according to extension), or spherical triangles from two opposite hemispheres, although in every respect which admits of being stated in terms intelligible to the mind through a verbal description they can be substituted for one another, there is yet a diversity which makes it impossible for the boundaries of extension to coincide. It is therefore clear that in these cases the diversity, that is the incongruence, cannot be apprehended except by pure intuition."

are must be differences in metrical properties. But the assumption is not supported, and without it, the argument only shows that deciding that differences in metrical properties make for geometrical dissimilarity leaves open the question of whether differences in chiral properties do as well, and the really interesting work starts when we try to answer the latter question in its own right. But Bennett et al. leave the assumption tacit and the interesting work undone.

6. PARITY VIOLATION

> I don't know whether anyone has written to you as yet . . . we are all rather shaken by the death of our beloved friend, parity. (John Blatt to Pauli, Jan. 15, 1957)

It is worth emphasizing that the question we are asking is whether *physical* objects are duplicates of their reflections, and hence whether reflection is a symmetry of *physical* space. We define mathematical spaces by starting with a manifold and defining thereon a set of geometrical objects; the only questions about the symmetries of such spaces are straightforward mathematical ones about whether the geometric objects in question are invariant under one or another transformation. When we ask questions about the structure of physical space, however, we are asking which of these mathematical spaces correctly represent it, and that is an empirical question. To say that T is a symmetry of a space, S, is to say that T preserves all intrinsic geometric properties of figures in S, and to say that T preserves all intrinsic geometric properties of S-figures is to say that anything that is *not* invariant under T is *not* an intrinsic geometric property of such figures. The claim that translation is a symmetry, for example, is the claim that all intrinsic geometric properties of physical objects are preserved by translation, and this just means we don't change an object internally by moving it. The question of whether reflection is a symmetry of physical space is the question of whether objects are altered internally by reflection.

We used to think the world was like a great big watch, and the job of the scientist was to explain the motion of the hands on its face by describing the hidden wheels and cogs which turn them. The analogy turned out to be unsustainable because of the suggestion that the parts interact only mechanically, but for some purposes it is still apt. Just as the wheels that turn in an efficient watch are the hands on the face and the wheels inside needed to make it tick, in theorizing about the physical world we describe not only the observable motions of visible bodies but the goings-on behind the scenes which sustain them, and quite apart from the question of whether there is an observable difference between an object and its reflected counterpart, we might need to recognize an unobservable difference to explain differences in their observable behavior. If the behavior of an object depends ineliminably on its orientation, this may be the best reason we can have for recognizing intrinsic differences between left and right.

There is, of course, precedent for postulating unobservable spatiotemporal structure to explain differences in observable behavior; we attribute different 'absolute' states of motion to objects which exhibit the same relative motions to explain the differential effects of acceleration. That Newtonian Mechanics and Special Relativity retained the distinction was, notoriously, the source of Einstein's dissatisfaction with them, expressed here in a passage from "The Foundation of the General Theory of Relativity":

> Two fluid bodies of the same size and nature hover freely in space at so great a distance from each other and from all other masses that only those gravitational forces need to be taken into account which arise from the interaction of different parts of the same body. Let the distance between the two bodies be invariable, and in neither of the bodies let there be any relative movements of the parts with respect to one another. But let either mass, as judged by an observer at rest relatively to the other mass, rotate with constant angular velocity about the line joining the masses. This is a verifiable relative motion of the two bodies. Now let us imagine that each of the bodies has been surveyed by means of measuring instruments at rest relatively to itself, and let the surface of S_1 prove to be a sphere and that of S_2 an ellipsoid of revolution.
>
> Thereupon we put the question—What is the reason for this difference in the two bodies? . . .
>
> Newtonian mechanics does not give a satisfactory answer to this question. It pronounces as follows: —The laws of mechanics apply to the space R_1, in respect to which the body S_1 is at rest, but not the space R_2, in respect to which the body S_2 is at rest. But the privileged space R_1 of Galileo [an inertial frame] thus introduced, is a merely facti-

> tious cause, and not a thing that can be observed. It is therefore clear that Newton's mechanics does not really satisfy the requirement of causality in the case under consideration, but only apparently does so, since it makes the factitious cause R_1 responsible for the observable difference in the bodies S_1 and S_2. (*The Principle of Relativity*, New York: Dover, 1952, p. 112-113)

The example of the globes brings out the unobservable structure (the distinction between inertial and non-inertial motion) invoked in the Newtonian explanation of the differences in their observable behavior. The observable difference is that S_2 but not S_1 bulges at the equator, and the explanation is that S_2 but not S_1 is moving non-inertially, and hence S_2 but not S_1 experiences distorting centrifugal forces which cause it to bulge. As Einstein points out, there is nothing in the relative motions of the spheres which distinguishes the two, and he derides as illegitimate the imputation of a 'factitious' cause (the attribution of theoretically distinct but observationally indistinguishable states of motion to the two) to explain S's bulge. As he presents it, General Relativity was formulated precisely to redress this perceived defect. It rather notoriously fails to do so,[15] but the important point for our purposes is that nobody but Einstein and the positivists would hold this against the theory. Everybody but the positivists and Einstein himself (at least the time-slice of Einstein that penned this passage) regard this objection of his as overly scrupulous; the procedure of postulating unobservable structure to explain observable differences in behavior is part and parcel of scientific theorizing; it is what theorizing is all about.

There is another question which bears an even closer analogy to the one we have been discussing: that of whether dilation is a symmetry of physical space. We know that dilations preserve relative lengths, orientations, etc., of objects, and if we try to imagine worlds obtained from one another by dilation—one that doubles all lengths, for instance—we have divided intuitions about whether they really differ. There is the story about a young boy afraid to get on an airplane because he is convinced that because

---

15. So long as we agree that the only observable spatiotemporal structures are the relative positions of material bodies and their relative motions, both the Special *and* the General Theory of Relativity incorporate unobservable structure, for both recognize differences in the states of motion of inertial and accelerating systems, and hence neither is symmetric with respect to the full set of transformations which preserve relative motions.

they appear so small from the ground, they (and everything in them) shrink when they get high in the air. His father, having succeeded in getting the boy to agree to a trip to his grandmother, turns to him in the air with the triumphant remark 'you can see for yourself that nothing has shrunk'. The boy quite rightly replies that he can see no such thing!

We know, of course, that the laws of nature are not invariant under changes of scale, and so if everything did shrink as it ascended in airplanes, we would be able to tell; experiments performed in the air would have different results than they do on the ground. Galileo seems to have been the first one to recognize the fact, and he attached considerable importance to it, reporting it along with the discovery of his laws of motion in *Two New Sciences*. His own arguments had to do with the strength of rods and bones (he reasoned that if you need a bigger bone for a bigger animal, e.g., a horse twice as high, wide, and thick as some actual one, you would need a bone that could hold eight times the weight, but what a bone can hold depends on its cross-section and that is only quadrupled by doubling the size of the bone, making it fit to support only four times the weight of the original), but we could make the point nowadays by pointing to the atomic structure of matter, or the invariant velocity of light.

There might be similar kinds of physical reasons for denying that reflection is a symmetry, it might be, that is, that the natural laws are not invariant under reflection. Here are three ways of expressing the claim that the laws are invariant with respect to an exchange of left and right (they are equivalent, but I include all because they lend one another content):

(i) if a phenomenon is permitted by the laws, so is its mirror image,
(ii) if a movie of any physical process whatsoever is shown with the film right-left reversed, one would not be able to tell from *knowledge of laws alone* that it was reversed (this is not to say that knowledge of initial conditions wouldn't decide the case [if, for example, the film showed us from the inside with our hearts pumping on the right], the point is simply that there would be nothing in the reversed film that violated the laws [the fact that most of us have hearts

on the left is, as we understand it, a contingent fact, an artifact of initial conditions]),

(iii) given two experimental set-ups with initial states I_1 and I_2 and resultant states R_1 and R_2 (arising from these by deterministic evolution), if I1 is an enantiomorph of I_2, then R_1 is an enantiomorph of R_2, i.e., experiments whose initial states differ only in handedness have outcomes that differ only in handedness.

Up until the late 1950's it seemed pretty certain that the laws were invariant under reflection, or—to use the more common expression—that parity was conserved. So much so that in 1952 Hermann Weyl was able to write with complete confidence:

> ... in all physics nothing has shown up indicating an intrinsic difference of left and right. Just as all points and all directions in space are equivalent, so are left and right. Position, direction, left and right are relative concepts. (*Symmetry*, p. 20)

Not five years later, however, there were experimental developments that flew in the face of Weyl's confidence. In a 1956 conference on nuclear physics in Rochester, an experimental physicist named Martin Block was sharing a hotel room with Richard Feynman and suggested to him—only half seriously—that a certain puzzle about elementary particles might be easily resolved by giving up parity. In the next day's meeting Feynman raised the question prefacing it with a credit to Block, remarking later that people teased him "and said [his] prefacing remark was made because [he] was afraid to be associated with such a wild idea", revealing the complacency of the prevailing mood about parity at the time. Yang and Lee, a pair of theoretical physicists, were present at the session, one of them responded to Feynman's question, and later that year the two published a paper, "Question of Parity Conservation in Weak Interactions",[16] in which they argued that there was really no good evidence that parity is conserved in weak interactions, and making some experimental proposals to test the question. The paper aroused little immediate interest, as Freeman Dyson wrote in *Scientific American* article in Sept. 1958:

> A copy of [the Lee and Yang paper] was sent to me and I read it. I read it twice. I said, 'This is very interesting,' or words to that effect. But I had not the imagination to say, 'By golly, if this is true it opens up a

---

16. *Physical Review*, Oct. 1, 1956.

whole new branch of physics.' And I think other physicists, with very few exceptions, at that time were as unimaginative as I.

But among the handful of people whose excitement it *did* arouse was an experimental physicist at Columbia named Wu, who in 1956 performed a series of experiments establishing—astonishingly and as unequivocally as one could hope—that parity is in fact violated in weak interactions. The experiments involved the beta-decay of cobalt-60, a highly radioactive isotope of cobalt which continually emits electrons. A cobalt-60 nucleus can be imagined (falsely, but not in a way that matters here) as a tiny sphere that spins like a top on an axis labeled north and south at the ends to indicate the magnetic poles. Normally, the electrons emitted are shot out from both the north and south ends of the nuclei, and since the nuclei are usually pointing in all different directions, this means that electrons are emitted in all directions. If the cobalt is cooled to near absolute zero to reduce the motion of the molecules due to thermal fluctuations, however, and a powerful electromagnetic field is applied, most of the nuclei can be made to line up with their north ends in the same direction, and this means that most of the electrons will be emitted in the direction along which the north and south ends of the magnetic axes point, and we can compare the frequencies of these which 'spin' to the right and those which 'spin' to the left. If the laws are symmetric with respect to spatial reflection, i.e. if parity is conserved, these should be roughly equal. As a matter of fact they are not; the atoms decay significantly more often with a right-handed spin than with a left-handed spin; regularly, and with a fixed ratio.[17]

The argument from these experimental facts to the claim that there are intrinsic differences between left and right, goes in two steps;

(i) the regularity is lawlike and hence the laws are not invariant under reflection,
(ii) the fact that the laws are not invariant under reflection entails that systems obtained from one another by reflection differ intrinsically.

---

[17] The personal details related are taken from Gardner's *The Ambidextrous Universe*, as related to him in correspondence with Feynman.

Each of these steps warrants a full discussion. The second is an application of Curie's Principle, which is the subject of Essay 2. The first concerns the relationship between the symmetries of the laws and the kinds of *de facto asymmetries* they are compatible with, and it is the subject of Essay 4

Essay 4: Asymmetry

1. INTRODUCTION

Under what conditions does an effect that is asymmetric with respect to a transformation **T** suggest that the laws which govern it are asymmetric with respect to **T**? Particular instances of the question have been discussed extensively: are there actual phenomena that suggest that the laws are asymmetric with respect to spatial reflection? Is the evidently time-asymmetric character of the evolution of thermodynamic systems compatible with the time-symmetry of the (allegedly) underlying statistical mechanical laws? Do the actual temporal asymmetries of knowledge, causation, counterfactual dependence, explanation, etc., suggest that the fundamental laws of the actual world are asymmetric with respect to reflection in time? For any of these to be live questions, it must be true that not just any old **T**-asymmetric phenomenon entails that the laws are **T**-asymmetric, but some do. The question is which? And why?

I will argue that (i) the division of the features of the world into laws and initial conditions is essential to our ability to maintain that the laws are symmetric with respect to any transformation that is not a symmetry of the manifold as a whole, (ii) the claim that a transformation **T** is not a symmetry of the laws is logically independent of (neither entails nor is entailed by) the claim that **T** is or is not a symmetry of the *initial conditions*, but (iii) the *improbability* of a non-random distribution of initial conditions suggests that asymmetries that cannot be traced back to a random set of initial conditions are 'troublesome', i.e. suggest asymmetric laws.

The name of the game in trying to reduce thermodynamics to the time-symmetric classical laws, for instance, is that of finding some assumption of fact, that doesn't involve postulating a non-random distribution of initial conditions, and that combines with the classical laws to yield the asymmetric behavior. In the case of reflection in space, our inability to give such an explanation of the results of Wu's famous experiments explains our concession to the failure of parity.

2. DISTINCTION BETWEEN LAWS AND INITIAL CONDITIONS

Wigner remarks somewhere on the aptness of legal terminology in speaking of the equations of our physical theories; just as the laws imposed by a government limit, without determining, the behavior of its citizens, the physical laws constrain relationships between events while still allowing for much freedom. The domain of their freedom, i.e. the respects in which systems can differ from one another while yet behaving in accordance with the laws, are the initial conditions. The distinction between laws and initial conditions is absolutely essential to our ability to maintain that the laws are symmetric even with respect to such simple transformations as displacements in space, rotations in space, or displacements in time given the evident fact that the world itself is asymmetric with respect to all of these transformations. What happens today differs from what happens tomorrow, the configuration of bodies at any given location or in any direction in space differs from that in any other. If the world were *in all respects* determined by the laws, however, it would possess the full symmetry of the laws, and would necessarily look precisely the same from any given temporal or spatial perspective. The fact that it doesn't shows that there are features of the world undetermined by the laws, and it is these contingent features which distinguish different points and different directions in space from one another, the past from the future, and in general any pair of distinguishable events related by a transformation which is a symmetry of the laws.[1] So if we ask: how could a universe governed by equations with such a high degree of symmetry appear so *asymmetrical*? The answer is simple: *all the asymmetry arises from asymmetries in the initial conditions, either the initial state of the world or individual chance departures from the symmetries of the initial state on the way to its present one.*

3. RELATION BETWEEN THE ASYMMETRIES OF THE LAWS AND THE ASYMMETRIES OF INITIAL CONDITIONS

Imagine a child is asked to make a clay figure that is bilaterally symmetric, and then allowed to paint it any way she pleases. There are many ways of painting it that preserve the bilateral symmetry, but she chooses to paint the left side of the figure green with orange polkadots and the right side orange with green polka dots. In doing so, she destroys the bilateral symmetry, but it may be that the distribution of color on one side can be obtained from that on the other by replacing orange with green and green with orange; this will be the case if the colors are the same shade and the polka dots are in the right places. If we ignore differences in color the painted figure is bilaterally symmetric, and if we take differences in color into account it is not, but—since one side can be obtained from the other by reflecting it through a plane through its center and switching the colors—the painted figure, even taking differences in color into account, is symmetric with respect to the product of a reflection and a color inversion.

Suppose, now, that the figure is taken home in a knapsack and knocks around long enough to obtain scratches, chips, and cracks here and there, distributed randomly over the whole. It is no longer symmetric in either of these ways (the left half cannot be obtained from the right half, nor even from the right half color-inverted), but we can still discuss its geometrical symmetries or its symmetries in color, ignoring the odd scratch or knick, or just its color symmetries ignoring the odd scratch but no knicks, or... With respect to any set of its (intrinsic) properties we can speak of the figure's symmetries. These will not in general be the same for

---

1. By 'distinguishable' here, I mean 'not perfectly similar', i.e. distinguished by the value of some quantity, observable or not.

different groups of properties, and the symmetries of the figure for *all* properties considered together will be the intersection of the groups of symmetries of each. If there is some property—like the distribution of knicks and scratches—which is genuinely random, then the latter group will contain only the identity and any (large enough)[2] part of the figure will be distinguishable (by some property in the group) from any other.

Consider a history of the world, a complete catalogue of the values that the fundamental physical quantities take at each point in space-time. In precisely the way we could talk about the symmetries of the clay figure ignoring differences in color, we can talk about the symmetries of the manifold ignoring differences in initial conditions. These are the symmetries of the laws. Whatever differences there are between manifolds obtained from one another by a symmetry of the laws will be determined by a difference in initial conditions, just as in the analogy with the clay figure whatever differences there are between figures obtained by symmetries of color and shape are differences in the distribution of scratches and knicks. Furthermore, just as the symmetries of shape and color place no constraints on the distribution of knicks and scratches, the symmetries of the laws place no constraints whatsoever on the symmetries of the initial conditions.

So, what is the relationship between the symmetries of the laws and the symmetries of initial conditions? There is no interesting *logical* relationship. Given any set of laws **L** and any transformation **T**, there is a model of **L** in which **T** is a symmetry of the initial conditions (regardless of whether it is a symmetry of L). And given any transformation T' which is a symmetry of L, there are models of L in which it is not a symmetry of the initial conditions. Consider, for example, a world governed by deterministic dynamical laws, and suppose that the world contains just two real-valued physical quantities. The state at any time is characterized by an assignment of ordered pairs of real numbers to points in space-time, and these in turn are determined by an assignment of pairs to points at an initial state.[3] It is easy to see that for any transformation T, there will be both initial states which are and initial states which *are not* symmetric with respect to T.[4] This is so whether or

---

2. What counts as a large enough part depends on the density of knicks and scratches.
3. It is customary to take the 'first' state, though in a bideterministic theory, there is nothing in the relations of nomological determination between states to distinguish the first from any other.
4. So long as T is not the identity.

not the laws themselves are symmetric with respect to T; for the laws only dictate the evolution of a world once its initial state is given. If T is a displacement through some (non-zero) spatial interval, for example, the initial state which assigns the pair <2,2> to every point and that which assigns a different pair to every point (say, its coordinates in some Cartesian system of coordinates), are T-symmetric and T-asymmetric, respectively. The most that can be said is that so long as the laws are symmetric with respect to T, a world started in an initial T-asymmetric state will remain symmetric throughout its evolution; so, for example, the <2,2> world above cannot evolve into a state asymmetric with respect to displacements through d.[5] Beyond this, the symmetries of the initial conditions and the symmetries of the laws have nothing to do with one another.

4. INITIAL CONDITIONS

If the symmetries of the laws and those of the initial conditions are logically independent of one another, we can never *infer* from the existence of a T-asymmetric phenomenon alone that the laws which govern it are T-asymmetric; given the right initial conditions, phenomena exhibiting whatever sort of asymmetry you please can arise from T-symmetric laws. But we can guess that the laws governing certain phenomena are asymmetric by thinking not about what is possible, but about what is likely; in particular, by thinking about the overwhelming *unlikelihood* of the sorts of initial conditions that would be needed to give rise to them *via* symmetric laws. Here's what I mean: given a system of equations, we can define a standard measure over the set of its models. There is a model for every possible set of initial conditions, and the measure of the subset of models in which the initial conditions display any given correlation is overwhelmingly smaller than that of those which do not. It follows that it is overwhelmingly more likely that within a given world, any correlation arises as a result of a laws associating them than that the correlated events are among the initial conditions, and it follows from this that - in ignorance of the laws - inferring from the existence of a correlation to a law is a good bet. Once again, it is not an essential feature of initial conditions that they are random; on the contrary, given any set of

---

5. This should be recognizable, from the discussion in essay 2, as an instance of Curie's Principle.

laws, there are models in which the initial conditions exhibit any fantastic correlation you please. It is rather that significant correlations between initial conditions are overwhelmingly *unlikely*, so if T is an asymmetry in the history of the *actual* world which can't be derived by symmetric laws from asymmetries in a set of random initial conditions, it is overwhelmingly likely that the laws themselves are asymmetric with respect to T. This unlikelihood is the origin of the methodological injunction to deduce all apparent order from the laws.

Return to the analogy with the clay figure. Suppose that whether any given point lies along a scratch is an accident, unconstrained by the laws and independent of the existence of scratches at elsewhere. It would be an astounding coincidence if such a bunch of independent accidents conspired to produce anything but a random pattern; the larger the set of scratches and knicks, and the more ordered the pattern, the greater the unlikelihood. So if knicks and scratches occurred at perfectly regular intervals or formed a word or image of the face of Sister Theresa, it would be a pretty good bet that there had been some design in their placement. All of this will be clearer in the context of some more realistic scientific examples. I'll take up the case of reflection in space, and then the case of reflection in time where these issues come to a head in the treatment of thermodynamic phenomena.

5. REFLECTION IN SPACE

The question of whether the laws are invariant under reflection in space is the question of whether all left/right asymmetries that there in fact are can be derived from initial conditions which contain a random distribution of right and left-handed events. Let T be reflection in space and A some intrinsically right-handed event ($\sim(TA=A)$). If - as it happens - events of type A occur significantly more often than those of type TA, then so long as the laws are symmetric with respect to reflection, one of the following must be the case:

(i) the A's are chancy events and the predominance of A's over TA's is the result of a bunch of them happening to go one way rather than the other,

(ii) the A's are not themselves chancy but are causal products of chancy events. This allows for a spectrum of possibilities: at the left end of the spectrum all the A's are causal products of a single event; at the right, each of the A's has its causal origin in an event among the initial conditions that is independent of the origins of each of the others,
(iii) the A's are neither chancy nor the causal products of chancy events but are derived deterministically from asymmetries in the initial state of the universe. This too allows for a spectrum of possibilities; at the right end, each of the A's is the result of the same single event among the initial conditions, at the left, each is the result of a different one.

The overwhelming unlikelihood of the preponderance *among the initial conditions* of asymmetric events over their mirror images, makes equally improbable explanations too far towards the right end of the spectrums in (ii) and (iii). If these are the only available options, it is a much better bet that the laws are asymmetric with respect to reflection.

It's not difficult to make a list of facts that distinguish left and right: I write with my right hand, the keyboards of pianos are typically arranged so that to a person seated in the position of a player, the lower notes are on the left and the higher notes on the right, people's hearts are characteristically on the left side of their chests, their intestinal tracts wind to the right, and so on. The first of these can be written off to chance. The others require a bit more work but are easily enough explained without recognizing asymmetries in the physical laws: all of the instances of the regularity are products of a single event which could just as well have gone one way as the other. In the case of pianos, the original choice of the order of the keys is arbitrary, but the fact that we make new pianos in conformity with the old is obviously not. In the case of human hearts and intestines, some initial whim on the part of God put Adam's heart on the left and wound his intestines to the right, but the fact that both are inheritable traits explains how they were passed by Adam down through the generations, and why they are traits we all share.

A prima facie more troublesome left/right asymmetry was discovered by Louis Pasteur in 1848; most of the solid bodies in our surroundings are crystalline, i.e. their grains have a regular inner structure arising from the arrangement of the atoms in regular lattices. These are formed from one single crystal, or—more commonly—a conglomerate of microscopic crystals of various sizes. Crystalline and polycrystalline substances make up the greater part of all solid bodies found in nature; ice, sand and loam, for example, are crystalline; metals and rocks are almost all polycrystalline. Apart from the glasses and substances of organic origin like wood, there are very few non-crystalline substances.

Among the geometrically possible systems of crystal symmetry, there are some ('enantiomorph crystals') which do not have bilateral symmetry, and which crystallize from optically-active substances (i.e. substances which turn the plane of polarized light to the left or right, known respectively as 'laevo-active' and 'dextro-active'). One might expect, if the laws are symmetric under reflection, that if the laevo-form of such a substance exists the dextro-form exists as well and that the two occur with approximately equal frequency. What Pasteur found was that when the sodium ammonium salt of optically inactive racemic acid is recrystallized from an aqueous solution at a lower temperature the deposit consists of the laevo- and dextro-form of an enantiomorph crystal. The latter is the tartaric acid present in fermenting grapes, but—and this is the strange thing—the former *doesn't appear to occur naturally at all*. Moreover, Pasteur discovered, this is true of almost all asymmetric carbonic compounds: most of those which occur naturally do so in only one of their enantiomorphic forms. Pasteur was quick to conclude, from this and from the fact that the only known methods of isolating the laevo- and dextro- forms of such compounds at the time relied on the enzymatic action of biological organisms, that the source of the asymmetry lay in the asymmetric chemical constitution of living things.[6] Supposing that he was right, does it follow that the biological laws are asymmetric under reflection, or can the asymmetry be traced back to asymmetries in the initial conditions which determined the develop-

---

6. In fact, he went farther than this and argued that the ability to distinguish right and left is the prerogative of living things, and that it marked the single principled difference between the chemistry of dead and living matter. As it turns out, he was wrong, for we know nowadays of ways of activating inactive substances that don't use living matter.

ment of organic matter, can the asymmetry be traced, not improbably, back to accidents at the genesis of life?

One view about the origin of organic matter is that it is due to accidents which - once a certain stage of evolution is reached - are apt to occur independently now here, now there. If this were right, the asymmetric molecules found in plants and animals would have a number of independent causal origins and if the laws were symmetric under reflection, there would be a mystery about why their laevo- and dextro- forms fail to occur with roughly equal frequency. Pascual Jordan, in fact argued that this consideration argued strongly in favor of a common origin for all of organic life. An alternative to both of these might be to trace the asymmetry to some global asymmetry, one having to do with the rotation of the earth, the direction of light received from the sun, or some such thing. Yet another alternative is to suppose that development actually started from an equal distribution of enantiomorphic forms, but that this was an unstable equilibrium which tumbled over under a chance perturbation. Whatever the truth of the matter, insofar as the name of the game is preserving the symmetry of the physical laws with respect to spatial reflection, it is played by trying to explain *de facto* left/right asymmetries by tracing them to asymmetries in initial conditions *without attributing any left/right asymmetry to the initial conditions as a set.* Pasteur's discovery challenged, without really threatening our lead; in 1956, however, Wu performed the experiments described in the last essay, which ultimately spelled our defeat.

The details, of course, aren't important. What is important is that the asymmetry in the results is robust and significant enough that it cannot be written off as the coincidental collaboration of individually chancy events, and so strongly suggests the existence laws which are asymmetric with respect to spatial reflection, and hence the existence of a quantity or spatiotemporal structure that distinguishes left from right. What is especially puzzling about the business is that all of the laws which we already know about are symmetric with respect to spatial reflection; what is especially puzzling about it, that is to say, is that differences between right and left should show up in *only* in esoteric phenomena arising from the weak nuclear interaction.

6. REFLECTION IN TIME AND THERMODYNAMICS

Turning, now, from reflection in space to reflection in time, and to the question of whether there are phenomena which, by the criteria of section (iv), cannot be accommodated by laws which are symmetric with respect to reflection in time (for short, temporally symmetric laws). We will focus on thermal phenomena (those involved with heat and the relationship between heat and work), and ask whether the temporally asymmetric phenomena which fall under the Thermodynamic Laws can be accommodated by the temporally symmetric classical laws to which it has been proposed they should be reduced. I'll make two assumptions that will allow us to put the question in a sharp form and avoid (arguably, though I won't argue it) irrelevant complications:

(i) that thermodynamics is a phenomenologically correct theory and hence that the principles which are known as the Laws of Thermodynamics express true empirical generalizations (I'll refer to them from now on as 'postulates', so as not to assume they have a lawlike status), and

(ii) that the laws of classical mechanics are wholly true.

If we believe what our scientists nowadays say, the latter assumption, though standardly adopted, is almost certainly false; quantum mechanics (or some related theory) is the right theory about the way things behave at the microscopic level, but taking into account the non-classical features of the quantum world introduces complications which are not thought to bear on the issue at hand, and if they are indeed irrelevant, it is best to treat the problem in the classical setting and transpose it to the more complicated one afterward.[7] For our purposes, it doesn't matter whether

---

7. Some of the reasons given are (roughly) that: (i) the dynamical laws of quantum mechanics are temporally symmetric in the relevant sense, (ii) the irreversible processes described in the Second Postulate occur at a larger scale of magnitude than that at which quantum-mechanical considerations come into play, (iii) such irreversibility as the quantum mechanical laws possess is introduced by the Projection Postulate and hence applies only to systems undergoing measurements, but the temporal asymmetry encapsulated in the Second Postulate is supposed to apply even outside measurement contexts. (see Watanabe (1935), Schrodinger (1950), Rosenfeld (1955)

(ii) is the most persuasive; it is not obvious that (i) is even correct, and (iii) applies only to interpretations of quantum mechanics which incorporate a projection postulate.

the classical laws are false; we care only whether, and in what way, they conflict with the temporally asymmetric phenomena which fall into the domain of Thermodynamics.

(i) Thermal phenomena

Thermodynamics is the study of thermal phenomena, processes involving the transformation of heat into work and work into heat. As we now understand it, heat just is a form of energy (specifically, the kinetic energy of the molecules and atoms of which a substance is composed), but in the days in which scientists were beginning to formulate thermodynamics, they were laboring under a burden bequeathed them by the caloric theory, a conception of heat according to which it is a fluid that is passed from body to body when they are placed in contact. The misconception did not, however, hinder the experimental study of thermal phenomena, and by 1842 Carnot had arrived at a principle which was essentially the Second Postulate of Thermodynamics. Eighteen years later Mayer announced the First Postulate, and the pure theory was completed early in this century when Nernst proved the Third.

The First Postulate is just a statement of the principle of conservation of energy; it says that the variation of energy of a system during any transformation is equal to the amount of energy that the system receives from its environment. It allows for the transformation of any form of energy into any other (of heat into work and work into heat), so long as the total amount remains constant, i.e. so long as no energy is created or lost in the process. The Second Postulate places limits on the transformation of heat into work. Here, respectively, are Kelvin's and Claussius's formulations:

> (II) A transformation whose only final result is to transform into work heat extracted from a source which is at the same temperature throughout is impossible.
>
> (II) A transformation whose only final result is to transfer heat from a body at a given temperature to a body at a higher temperature is impossible.

Below, once we have introduced the relevant definitions, I'll give yet another formulation. The empirical content of the theory is entirely contained in the first Two Postulates; the Third is actually a theorem which describes a useful property of thermodynamic state-spaces (and for that reason, not quite appropriately referred to as either a postulate or a law, but I'll tolerate the awkwardness for the sake of complicity with custom), and I'll touch on it below. Our concern is almost entirely with the Second Postulate, because—unlike the First or Third, and indeed unlike any other laws that were known or suspected until quite recently—the Second Postulate is not temporally symmetric.[8]

(ii) Thermodynamic states

The dynamical states of classical systems consist of the position and velocity of each of the particles of the system's constituent particles; for a system consisting of n particles, it takes 6n numbers to specify these (three for position coordinates and three for velocity components of each particle). In dealing with big systems, systems for which n is very large, it is obviously impracticable to try to determine and work with their dynamical states, but it is not always necessary, for much of the precise information contained in a system's dynamical state is irrelevant to some of its properties. When we are lucky, we can fix on 'higher-level' laws or generalizations which capture all of the regular behavior of the properties we happen to be interested in, without needing to advert to distinctions between dynamical states to which these are indifferent.[9] Something like this is true in the case of thermal phenomena; the only parameters needed to describe them, or, as it turns out, to fix on generalizations which express the regularities in their behavior, are the temperature (T), pressure (P), and volume (V) of the systems involved. And these are relatively coarse grained properties in the sense that they are, to a very great degree, independent of

---

8. Now, of course, we have quantum mechanics and whether, and in what sense, the laws of quantum mechanics are symmetric with respect to temporal reflection, depends (in part) on undecided questions about the interpretation of the theory.
9. In saying that if we are lucky we will be able to fix on higher level laws or generalizations which capture all of the regular behavior of the systems of interest, I don't mean merely that there exists somewhere in logical space a set of sentences relating only the properties of interest which correctly describes the regularities they exhibit. There will always be that. I mean that the regularities in the pattern they exhibit will emerge clearly without consideration of the richer pattern in which they are embedded, in the way that the pattern of drum beats in an orchestral work is easier to map without considering the piece in which they are embedded than, say, the relations between two notes of a given pitch in the first trumpet's part.

the exact positions and velocities of the particles comprising them; a large number of microscopic arrangements may underlie a single set of values for P, V, and T.

The state-dependent properties of a thermodynamic system are its values for P, V, and T, and the three are not independent of one another; they are related by an equation of state $f(P,V,T)=0$ whose specific form (where what exactly f is), depends on the type of the system in question (its chemical composition, etc.). This means that the values of any two of the parameters fixes the third, and it allows us to represent the states of a thermodynamic system in a state-space of just two dimensions. Given a system S whose equation of state is $f_s(P,V,T)=0$, for example, we can construct its state-space by plotting V along one axis in a plane and values of P along the other. Any point in the V,P-plane will correspond to an assignment of values to P and V, and since these determine a value for T by f_s, every physically possible state will correspond to some point in the plane.

(iii) Entropy

Any function of the thermodynamic state of a system is a thermodynamic quantity, and here I'll define a quantity, the entropy, which will allow us to make the above remarks about the relation between thermodynamic states and classical dynamical states precise.[10] First, some preliminary definitions; *transformations* are transitions between initial and final states of a system through a continuous series of intervening ones. If the states of a system are represented on a V, P-diagram, a transformation will correspond to a curve connecting points representing the initial and final states and the points through which the curve passes will represent intervening states. *States of equilibrium* are states with the property that they don't change so long as external conditions remain constant, and *reversible transformations* are transformations whose successive states differ only infinitesimally from equilibrium states.

Now, if we consider two equilibrium states, A and B, for a system represented on a V,P diagram, we can show that the value of

---

10. Entropy is a function of the state of a system; it is indifferent whether we regard it as a property of states or as a property of systems. It is more customary to do the former, but the latter accords better with what I say about physical properties in essay 5, section 5.7.

the integral $\int_B^A dQ/T$ along any reversible path from B to A is the same. This means we can define a function $S_O(A)=\int_O^A dQ/T$ by fixing a standard state O, and taking the integral along any reversible path from O to A.[11] Functions defined by making different choices for O will in general differ from one another by a constant, and this means that the *difference* between the S-values of any pair of states is indifferent to the choice of standard state (i.e., $S_P(A)-S_P(B)=S_Q(A)-S_Q(B)$, for any choice of P and Q), but the absolute values assigned each are not (i.e., it is not in general true that $S_P(A)=S_Q(A)$). By proving that all states take the same value at T=0 for any of the S-functions, i.e. that $S_O(A)-S_O(B)=0$ for all states A and B at T=0, however, Nernst showed that the arbitrariness in the choice of a standard state can be removed by taking an equilibrium state at T=0, since the function defined thereby is indifferent to which of *these* is chosen.[12] So we define a quantity, the 'entropy' of a system, as the function of its thermodynamic state A, given by $S(A)= \int_{t=0}^A dQ/T$ in prose, the entropy of a state A is the integral of the amount of heat absorbed by a system in A from another system at a temperature T in any sequence of reversible transformations that take it from an equilibrium state at T=0 to A, over T.

So, the procedure used to define entropy is this: we notice first that the value of the integral $\int_B^A dQ/T$ for a reversible transformation depends only on the initial and final states of the transformation, A and B, and so if we let $\int_B^A dQ/T$ be the integral taken along any reversible path from A to B, we can define a family of functions of the state $S_O(A)=\int_O^A dQ/T$ by making different choices for O. It turns out that in general the functions in this family differ by a constant, but each assigns the same value to all states at T=0, and this latter fact is exploited to reduce the arbitrariness in the choice of the standard state, so the entropy of a state A, is defined as the value of $\int_{t=0}^A dQ/T$ taken along any reversible path.

Now we can state the prohibition expressed by the Second Postulate of Thermodynamics quite simply as a ban on transforma-

---

11. O and A must both be equilibrium states, for there will only be a reversible path between them if they are.

12. Or, rather, he showed that no such arbitrary choice needs to be made by showing that all states have the same entropy at T=0.

13. We built it into the definition that the difference in entropy between states A and B, $S(A)-S(B)= \int_B^A dQ/T$ if the integral is taken over a reversible path from A to B, but this leaves open the question of whether the equation holds if the integral is taken over a path which contains irreversible transformations. The Second Postulate allows that the entropy increases in such cases, $S(A)-S(B)= \int_B^A dQ/T$, but states that it never decreases.

tions which take isolated systems into states with lower entropy than their initial states, for short, a ban on entropy-decreasing processes.[13] If A is the initial state and B any subsequent state of any real isolated system, then $S(A) \leq S(B)$; if the transformation is reversible, $S(A)=S(B)$; if it is irreversible, $S(A)<S(B)$. It should be emphasized that the prohibition on entropy decreasing processes applies only to isolated systems. It is always possible to reduce the entropy of a system (provided, of course, it is non-zero) by coupling it with a system in a lower-entropy state; the Second Postulate entails only that in such cases, the entropy of the latter is accompanied by a decrease in that of the former, and the overall entropy of the pair either increases or (if the transformation is a reversible one) remains constant.

(iv) Statistical mechanics

This quantity plays an important role in making precise the remarks I made about thermodynamic states being coarse-grained versions of classical dynamical states. I'll follow the custom of calling the former 'macrostates' and the latter 'microstates' from now on. The microstate of an n-particle system, as I mentioned, is given by 6n numbers and can be represented by a point in a 6n-dimensional phase space. Instead of the precise characterization of the microstate given by its phase-space coordinates, the microstate can be characterized approximately by dividing the phase-space into cells of hypervolume t, and specifying only the cell into which the microstate in question falls.[14] In this way we substitute for the continuous set of perfectly precise alternatives corresponding to different microstates, a discrete set of imprecise alternatives which can be counted up, added, subtracted and compared by the ordinary combinatory methods. When I speak below about the numbers of microstates compatible with different macrostates, what I really mean (though I won't continue to say so) is the numbers of cells of finite volume compatible with those states.[15]

---

14. I have allowed myself (for example, when I speak of the phase-space coordinates of a state and refer to the cell in phase-space into which a given state falls) to conflate states with the phase-space points which represent them where I didn't think any confusion would result.
15. I am deliberately glossing over issues concerning the arbitrariness introduced by the freedom in choosing the volume t, the way in which quantum mechanics dictates the choice, how Nernst's theorem suggests that quantum mechanics is correct on the point, because these are peripheral to our purposes.

The interesting fact about the relation between macrostates and microstates is not only that the former are coarse-grained versions of the latter, but that the coarse-graining is *uneven* in the sense that different macrostates may correspond to radically different numbers of microstates. It was Boltzmann who showed this, and he who made the relation macrostates and their corresponding microstates precise when he proved that the entropy of a macrostate is related to the number π of microstates compatible with it by the equation:

$$S = k \log p$$

where k is the ratio R/A where R is the gas constant and A is Avogadro's number.

What this means, and what it was Boltzmann's triumph to see, was that of all the microstates compatible with a given macrostate M, the number of those which evolve (in accordance with the classical dynamical laws) into states with *lower* entropy than M is so overwhelmingly dwarfed by the number of those that evolve into states of similar or higher entropy that, if we assume that all microstates are equally probable, we can expect entropy decreasing processes to occur *with vanishing probability*. In Schrodinger's words: if we say that all states evolve into states with similar or higher entropy, "we disregard only a very small fraction of all [those that are possible] - and this has 'a vanishing likelihood of ever being realized'."[16] So, Boltzmann proved his equation, added to it the assumption that all microstates are equally probable (the Equiprobability Assumption, as it is known), and concluded that entropy decreasing processes, although not impossible, have only a vanishing probability of occurring.

It is crucial in understanding Boltzmann's account, to be clear on the nature of the probabilities in question. Boltzmann supposed (and we are supposing with him) that the classical mechanical laws are wholly true, and we know that in classical mechanics every isolated system has perfectly definite *micro*state and its evolution throughout all time is proscribed by bideterministic dynamical laws, that is to say, its microstate at any time determines its microstate at all others. The probabilities that we are talking about are purely epistemic, they arise because if all we know about a sys-

---

16. Statistical Thermodynamics, Dover: New York, 1989, p. 6.

tem is its macrostate, then we are missing information about its microstate. The higher the entropy of the relevant macrostate, the more information is missing, for the higher the entropy, the greater the number of different microstates with which it is compatible, and the greater the number of different microstates with which it is compatible, the greater the degree of uncertainty as to which it is in fact in (again, assuming their equiprobability).

Consider a system S—a gas, say, of known chemical composition in an enclosed container—and suppose that we have measured the pressure and volume of S and so we know its macrostate M*. If we want to calculate the probability with which S, left to its own devices, will evolve into a lower entropy macrostate, we count up the number of microstates compatible with M* which the classical laws take into lower entropy states, and divide by the total number p of microstates compatible with M*. As it turns out, the probability is vanishingly small no matter what state M* is; for any macrostate whatsoever, the probability with which a system in that macrostate will evolve into a macrostate of lower entropy is as good as zero. The truth in the Second Postulate, on this understanding, is that whatever macrostate a system is in, the odds are overwhelmingly against its being in one of the ever so few microstates compatible with that macrostate which evolve into lower entropy states.[17] Such microstates are so overwhelmingly atypical of those compatible with any macrostate, that so long as our knowledge of the state of a system is limited to its macrostate, we are almost certain never to lose if we always bet against decreases in entropy.

The account isn't yet complete because there is a twist which Boltzmann didn't see, until some arguments of Loschmidt's brought it home to him. When he proved that the entropy of a macrostate is proportional to the log of the number of microstates compatible with it, Boltzmann did show, as he had wanted to, that the Second Postulate could be derived via the classical mechanical laws from the assumption of the equiprobability of all microstates,

---

17. On the contrary, we can show that there exist such microstates. If Boltzmann were right, and this were the whole story, and if we had - moreover - the right sort of control over microstates, we could prepare a system in such a state and actually observe a violation of (II). Unfortunately, of course, we don't have the requisite kind of control over microstates, and there might at any rate be reasons (e.g., those that Albert marshals) for thinking that even if Boltzmann's story were mostly correct, there are additional considerations which tell against the possibility of such a test.

but unfortunately, he also showed that the temporal reverse of the Second Postulate could be derived therefrom. Call an 'entropy-decreasing process' any process by which an isolated system evolves into a final state that has a lower entropy than that of the state in which it began, and call any process which takes a system in a given macrostate into one of higher entropy an 'entropy-increasing process'. The Second Postulate prohibits entropy-decreasing processes, its temporal inverse prohibits entropy-*in*creasing processes, and what Loschmidt is now famous for pointing out to Boltzmann is that the Equiprobability Assumption plus the classical dynamical laws not only entail that entropy-*de*creasing processes occur with vanishing probability, but also that entropy increasing processes do as well. This should have been clear from the start, for the assumption that all microstates are equiprobable is essentially the assumption that the universe as a whole is in a state of maximum entropy because the overwhelming majority of all microstates correspond to maximum entropy macrostates. And we know from the classical dynamical laws that if the universe as a whole is in a state of maximum entropy, it is as good as certain to stay there.[18] This was quite enough to show that those assumptions couldn't be correct, for the latter is flatly contradicted by experience: entropy increasing processes are always, actually occurring, everywhere.

The assumption that Boltzmann really wanted was one that would underwrite the application of the epistemic probabilities without dictating any particular macrostate for the universe. The assumption that he really wanted was that any microstate compatible with a given macrostate is as probable as any other microstate *compatible with that macrostate*, i.e., that for any macrostate A and any two microstates a1 and a2 compatible with A, $p(a_1)=p(a_2)$. Put another way, if we collect into a large ensemble all of the (actual and possible) systems in A, the various microstates $\{a_1, a_2...\}$ compatible with A occur with frequencies that approach equality as the size of the ensemble goes to infinity. If we restrict our attention to

---

[18]. This is not to say that it will not evolve, its changing microstate will be describing a path through (classical dynamical) phase-space proscribed by the classical laws. We obtain the result that its macrostate is unlikely to change by showing (roughly) that in the long run any path can be treated as a random walk, and then showing that a random walk starting at a point within the region corresponding to maximum entropy spends (very, very close to) no time outside that region.

the microstates compatible with a particular macrostate (low entropy states and high entropy states, alike), we can show two things; (i) that the overwhelming majority of these evolved—in accordance with the classical dynamical laws—*from* states of equal or lower entropy, and (ii) that the overwhelming majority of these will evolve—as proscribed by those laws—*into* states of equal or higher entropy.[19] The upshot, if we make the revised Equiprobability Assumption, is that any system—no matter what its macrostate—is overwhelmingly likely to be evolving towards maximum entropy, and away from a state of equal or lower entropy, indeed those which don't have a vanishing probability of occurrence.

If all of this is right, the prevalence of isolated or quasi-isolated systems ('branch systems', as Reichenbach called them) in the actual world which are evidently currently in states of non-maximum entropy, strongly suggests that the universe as a whole (or at least our local pocket of it) is presently in a low entropy state, and if this is right, then the only likely scenario for its history is that it started in an even lower entropy state and has been evolving towards maximum entropy ever since.

I am leaving out too many details and riding over too many niceties to mention, but in outline Boltzmann's account combines two assumptions of contingent fact with the classical dynamical laws to derive both that entropy-decreasing processes never occur *and* that entropy increasing ones very often do. The two assumptions of fact are (i) the Equiprobability Assumption (in its revised form), and (ii) the assumption that the initial state of the universe was one of low entropy. Boltzmann's account, that is to say, combines the Equiprobability assumption and the assumption of a low entropy initial state with the classical dynamical laws to derive both (a slightly weakened version of) the Second Postulate and the negation of its temporal reverse.

In section 3, I said that there is no logical incompatibility between T-symmetric laws and the world exhibiting asymmetries of any sort whatsoever, and it follows that there is no logical incompatibility between the temporal asymmetry of the classical dynamical laws and the temporal asymmetry of thermal phenom-

---

19. The 'equal' applies only if the macrostate is one of maximum entropy; for all others, we can show that the overwhelming majority of compatible microstates came from states of lower entropy and will evolve into higher entropy states.

ena encapsulated in the Second Postulate. So what exactly was Boltzmann's account answering to, what fact was it intended to explain or what worry was it intended to allay? Nobody doubted that the phenomena described in the Postulates of Thermodynamics were consistent with the classical dynamical laws, so the worry wasn't that the *negation* of the Second Postulate was entailed by those. And it was known from the start that reducing the Thermodynamic Postulates to classical mechanics was going to involve a weakening of the laws; all it takes to see this is to notice that the former are, and the latter are not, asymmetric with respect to temporal reflection. Boltzmann's account was meant to allay a different worry, one that was much more subtle; the worry was that any reduction would render the laws *too* weak.

The classical dynamical laws are temporally symmetric, so there are choices of initial conditions which would give rise to the temporal inverse of this world, a world in which there were only entropy-decreasing processes. The fact that the actual world contains only entropy increasing processes is an artifact of a choice of initial conditions which biased subsequent history in favor of entropy increasing processes. It just happened that in the beginning when things were being set up and there was a choice to be made between conditions that would give rise to an entropy increasing-process and those that would give rise to an entropy-decreasing process, it always went in favor of the entropy increasing process. There would be nothing strange in this if there were just a few choices to be made...but if there were many of choices to be made, and they were made completely independently of one another, we would be building into the very beginning of our account of how the present state of the world came about, an immensely improbable set of coincidences. That, as I put it in section 4, was the rationale for objecting to assumptions of fact which involve postulating significant departures from randomness in the initial conditions of the actual world. It was a fairly shallow rationale. A deeper rationale, one which anticipates some of the themes of the next essay, is that the criteria of section 4 are derivative of global criteria that apply to theories as wholes, and which have to do with the way in which we divide the features of the actual world

---

20. All of this is done in the language of the theory; the laws are given in that language and the initial conditions described in it. We look not for correlations in the 'bare phenomena', but between the theoretical quantities in terms of which the theory describes the features of the world it regards as initial conditions.

into laws and initial conditions. The division between the lawlike and the chancy should coincide, as closely as possible, to the division between regularity and randomness.[20] To put it in a picturesque way: the laws should cast their net widely enough, and be meshed closely enough, to capture all actual regularity. We tell if they are closely enough meshed precisely by examining the debris they leave behind, by looking for regularity in the initial conditions.[21]

(v) Conclusion

I started by articulating the conditions under which a T-asymmetric phenomenon can be accommodated by T-symmetric laws, *viz.* when it can be derived by way of them from random (or close to random) initial conditions. Then I applied it briefly to the case in which T = reflection in space, and less casually to the historically important case in which T = temporal reflection and the phenomena in question were those which fall under the Second Postulate of Thermodynamics. I suggested that the adequacy of proposed reductions to classical mechanics hinges on our ability to derive the Second Postulate therefrom without making objectionable assumptions of contingent fact, where an assumption of contingent fact is just an assumption about the nature of initial conditions in

---

21. Notice how this goes against the view expressed by Horwich in the following passage (and by Reichenbach, Grunbaum, Gold, Davies, and others, elsewhere), that a complete account would provide some justification for, or explanation of, the Equiprobability Assumption.

> "The theory is certainly quite plausible and attractive. However, there are a couple of respects in which it falls short of being a fully adequate account of one-way processes. In the first place, the fact concerning the commonness of low-starting-entropy branch systems, though evidently true, does not seem quite fundamental enough to constitute the basis of our explanation. We would like to know why it obtains. In the second place, the 'random initial microstate' hypothesis also requires explanation. Moreover, unlike the other assumption, it is far from obvious. So it would be desirable to find some independent reason for believing it."
> (Asymmetries in Time, p. 70)

If what I have said is correct, this is a mistake, for the empirical content of the equiprobability assumption is that the distribution of microstates compatible with a given macrostate is random, and it is exactly what we would expect if the initial macrostate of the universe were constrained but its microstate weren't (or weren't *otherwise*). Since these are just coarse-grainings of parameters which characterize the micro-state, indifferent to distinctions among microstates compatible with them, constraints on the macro-state alone should leave us with a random distribution of such microstates as are compatible with it. It is, rather, departures from randomness in the distribution of microstates that demand explanation.

the actual world and is objectionable to the extent that it forces a recognition of correlations among initial conditions. The shallow rationale I gave was that any particular departure from randomness is overwhelmingly improbable (where the probability is given by a standard measure over the theory's models defined relative to the partition of these defined by the theory's basic quantities). The deeper rationale was that we *make* the division between laws and initial conditions so that it coincides as closely as possible with that between regularity and randomness in the actual world, and that is a nice introduction to the topic of the next essay.[22]

---

22. I should say a word about the status of this talk about how laws are separated from initial conditions and about the methodological principles which govern theoretical choices. Is it descriptive or normative? Somewhere in between. The procedure is something like this: we begin with a description of a set of theoretical choices that are actually made and try to read off from these the extra-empirical standards against which theories are measured, i.e. the methodological criteria which guide choices between theoretical alternatives when the evidence itself doesn't tell between them. Here the kinds of idealized reconstructions of the history of their subject that scientists pass on to their students through textbook accounts, and the like, are a better guide than an historically accurate story sensitive to the real complexities that determine course of scientific development, for it is only in the idealized reconstructions that a set of clearly defined alternatives is presented, and the rationale for theoretical choices that are actually made and endorsed in retrospect is made explicit.

Essay 5
Science and Symmetry

1. INTRODUCTION

When we look at physical theories from the perspective of their symmetries what we see is the structure of their models, independently of their relation to the physical world. One question for which we might expect the perspective to be revealing is that of the theoretical basis for the choice between empirically equivalent theories, for it is only in the structure of their models that such theories differ. In this chapter, I take up this question from the viewpoint of symmetry and explore the answer that emerges. What we see is that the movement in theorizing (where not prompted by concern for empirical adequacy) is always in the direction of greater symmetry of models. This fact provides a basis for an account of scientific theorizing which differs in interesting ways from those which dominate the philosophical literature.

Before I get to the account itself, I will go over some theoretical background. First, I'll say something about the conception of theories with which I am operating, and clarify this notion of empirical equivalence between theories and the distinction between the observational and theoretical content of a theory on which it depends. Then I will say something about the content of scientific theories and address some of the traditional questions about the relation between our theories and the evidence. Much of this is old ground, but it is worth going over because the ultimate destination is unfamiliar and it will be well to keep track of the route by which we arrive, and to be able to situate it with respect to some familiar landmarks.

2. STRUCTURE OF THEORIES

Scientific theories are the products, variously presented and loosely construed, of science. Those of us who aren't scientists have more or less vague notions of what specific theories look like. More likely than not, we associate them with the quantitative theories familiar in the mathematical sciences, and know the names of a few, e.g., the general theory of relativity, evolution, and perhaps quantum mechanics. For the purposes of comparison and discussion in a philosophical context, we need a uniform and convenient way of representing arbitrary theories, one that is as precise as possible while being general enough to apply to all recognized cases. This, I take it, is what it is to give an account of theories. We have an intuitive general conception, a set of recognized instances, and some more or less vague notion of its boundaries, and we want to capture the general idea by providing a schema which brings out what the recognized cases share, and draws the boundaries somewhere within the range of vagueness allowed by the intuitive conception. There can be questions about the adequacy of various accounts (do they individuate theories in the right way? do they draw the boundaries in the right place?), and there can be pragmatic reasons for preferring one over another (one may bring the important features of theories sharply into relief and facilitate important comparisons, while another all but buries the important properties and brings out incidental features), but questions about which among the alternative adequate accounts is correct, are moot.

I'll introduce the two most influential conceptions of theories, the semantic conception and its chief rival, the syntactic conception, and argue that they are both adequate but that there are pragmatic reasons for preferring the former. In the Introduction I suggested a particular version of the semantic conception in terms of which the notions that occupied subsequent essays were most easily understood, but the points I want to make here are general and I won't distinguish among the various versions of the semantic conception.

(a) semantic and syntactic conceptions of theories

On the **semantic conception**, a theory is identified with a set of mathematical structures (its 'models') together with an interpretation, i.e. a mapping of elements and relations in the model onto physical elements and relations. More or less complex constructions of basic elements and relations are mapped onto macroscopic objects and their observable properties; these comprise the empirical substructures of a theory's models, and the structures in which they are embedded constitute their higher-level structure. A theory is empirically adequate just in case the actual configuration of macroscopic objects and their observable properties is isomorphic to the empirical substructures in one of its models, and a theory is true just in case the configuration of basic physical elements and their properties in the actual world is isomorphic to the higher-level structure of one of its models.[1]

On the **syntactic conception**, by contrast, a theory is represented as a deductively closed set of sentences in a canonical language L, where

(i) L is a first-order language with identity,
(ii) the non-logical terms of L are divided into three disjoint classes:
  (a) logical vocabulary consisting of logical constants (including mathematical terms),
  (b) observation vocabulary, V_o, containing observational terms, and
  (c) theoretical vocabulary, V_T, containing theoretical terms, and where
(iii) the terms in V_o are interpreted as referring to observable physical objects or observable attributes of physical objects.[2]

The term 'interpretation' has a technical meaning, here; an **interpretation** of L is a specification of a domain and assignment of individuals to names and sets of individuals to predicates, and L's

---

[1]. I will always, unless otherwise indicated, mean a claim about the representational or empirical adequacy of a model to be relative to the intended interpretation of the theory to which it belongs.

intended interpretation is that which gives the intended meanings of L-terms, as they figure in T. T is **empirically adequate** just in case the theorems containing only observational vocabulary are true under the intended interpretation, and it is true just in case *all* of its theorems are true thereunder.

This isn't, and wasn't presented as, a description of the way theories are actually presented and discussed in scientific contexts. Even the most cursory look at formulations in scientific journals and textbooks will reveal that theories are almost never formulated axiomatically, and they are certainly not presented in languages satisfying (i)-(iii). One sometimes finds this cited as a criticism of the account, but those who developed the account were not so unsubtle as to suppose that they were offering a description; it was put forward, rather, as what they called an 'explication', i.e., a precise and explicit rendering of a more or less vague or indeterminate concept. As Carnap writes:

> The task of explication consists in transforming a given more or less inexact concept into an exact one, or rather, in replacing the first by the second. We call the given concept (or the term used for it) the *explicandum*, and the exact concept proposed to take the place of the first (or the term proposed for it) the *explicatum* ... the explicatum must be given by explicit rules for its use, for example, by a definition which incorporates it into a well-constructed system of scientific either logico-mathematical or empirical concepts (Carnap, *Logical Foundations of Probability*, Chicago, Ill.: University of Chicago Press, 1950).

---

2. The positivists added the following conditions which embody the distinctive philosophical part of their view: (iv) the terms in VT are given an explicit definition in terms of Vo by correspondence rules C - that is, for every term 'F' in VT, there must be a definition of the form: (x)(Fx=Ox), where 'Ox' contains only logical vocabulary and symbols from Vo. The correspondence rules, at once define theoretical terms and specify operational procedures for applying a theory to phenomena.
(v) there is a set of theoretical postulates T whose only non-logical terms are from VT, and which encapsulate the specifically theoretical part of the theory.
It follows from this that any set S of sentences of L is equivalent to another set S' completely devoid of such terms, specifically, the set of sentences obtained by replacing each of the theoretical terms Fx in S by its definition in observational vocabulary. This means, in particular, that the content of any scientific theory can be given by a set of sentences containing only observational vocabulary, and hence that any two theories which are observationally equivalent are equivalent tout court.
A number of changes and refinements were made in the account, but the basic idea remained the same: a theory can be given a canonical formulation as a deductively closed set of sentences in a first-order language, and its content can be stated in purely observational terms.

Early proponents of the semantic conception of theories, perhaps because the syntactic conception had exercised such influence for so long, went about overthrowing it with a kind of revolutionary fervor.³ Once the dust had settled, however, it was difficult to see what exactly the charge against it was supposed to have been. If it was that the syntactic conception is the source of mistakes, i.e. that it encouraged some mistaken ideas, then it is correct. If it was that the syntactic conception is itself mistaken, then it is misplaced, for the two views are (with a relatively unimportant exception) equivalent; they count (almost) the same things theories and individuate them in the same way.⁴

(b) formal relations between them

A **model-in-the-logician's-sense** of a theory T is a structure which, under some interpretation (i.e. a specification of a domain and an assignment of elements in the domain to non-logical vocab-

---

3. Fred Suppe writes in the "Afterward" for *The Structure of Scientific Theories*:

    "For over thirty years logical positivism (or logical empiricism as it later came to be called) exerted near total dominance over the philosophy of science.
    Only in the 1960's did logical positivism's program for the philosophy of science come under serious intensive challenge ... the most fundamental and damaging were those aimed at the [syntactic conception of theories]. By the end of the decade these had been so successful that most philosophers of science had repudiated the [syntactic conception]." (p. 617)

4. Most likely, the two weren't carefully distinguished. Friedman, reads van Fraassen's attack on the syntactic conception as urging the incorrectness of the view, and takes him - rather sharply - to task:

    "Van Fraassen opposes his (semantic) account of empirical adequacy to the (syntactic) 'language-oriented' account characteristic of positivist philosophy of science...
    I find this anti-linguistic (anti-syntactic) attitude extremely puzzling....[in case the class of models of a theory is an elementary class] the Completeness Theorem immediately yields the equivalence of van Fraassen's account and the traditional syntactic account....
    [van Frassen] appears to be surprisingly unclear about the basic - and interesting - relationships between semantic and syntactic concepts". (Friedman, "Review of *The Scientific Image*," *J.Phil.*, 1982)

ulary occurring in T [objects and sets of n-tuples of objects to names and n-place predicates, respectively]), satisfies T. To say that the world provides the makings of a model of T is to say that the world is the domain of some interpretation which satisfies T.

A **model-in-the-sense-relevant-to-the-semantic-view**, by contrast, is a structure which, under a specified interpretation, represents a world that is physically possible according to T, under a specified interpretation. These are *not* the same thing. The former is a structure of which T, properly interpreted, could be true; the second assumes an interpretation, and is a structure of which T, so interpreted, *is* true. There are interesting relations between the structures in the class of **models-in-the-logicians'-sense** of a (first order) theory T and its class of **models-in-the-sense-relevant-to-the-semantic-view**.

If L is the set of T-models-in-the-logician's-sense, and M is its set of T-models-in-the-sense-of-the-semantic-view, then each structure in M is isomorphic to some structure in L. If, moreover, M is an elementary class, then the converse is true as well: L and M coincide. If, on the other hand, M is not an elementary class, there will be models in L which are not isomorphic to any in M. This happens only when T says that there are exactly denumerably many things, or says that there are at least uncountably many things.[5] So, one cannot always recover from the syntactic formulation of a theory, exactly the set of structures of which under a specified interpretation, it is true, but this is not a consideration which decisively favors either conception of theories, for several reasons. First, it is not obvious that our pre-philosophical notion

---

But much of what van Fraassen says suggests it is the pragmatic superiority of the semantic view that he means to be pushing. This is quite compatible with its equivalence (or near equivalence) to the syntactic view, and indeed quite compatible with full awareness of the interesting relationships between semantic and syntactic concepts. He writes in *The Scientific Image*:

"Impressed by the achievements of logic and foundational studies in mathematics at the beginning of this century, philosophers began to think of scientific theories in a language-oriented way.... Everyone knew that this was not a very faithful picture of how scientists do present theories, but held that it was a 'logical snapshot'...

A picture is only a picture - something to guide the imagination as we go along. I have proposed a new picture, still quite shallow, to guide the discussion of the most general features of scientific theories." (p. 64)

of theory decides the cases over which the two disagree. Proponents of the syntactic view can bite the bullet and say—with some, but not an intolerable amount of, awkwardness—that it is impossible for a theory to say either that there are exactly denumerably many or non-denumerably many physical objects. Second, even if we allow that it can be part of the content of a theory that the world has one or another infinite cardinality, a part of the content that can't be represented by any set of first-order sentences, there is room for proponents of the syntactic view to retrench. If T is formalized in a higher-order language, we can often get a closer fit between the set of T-models-in-the-sense-of-the-semantic-view and T-models-in-logician's-sense. There are pragmatic reasons for favoring formalizations in first order languages (specifically, all first order theories, but not all second-order theories are decidable), but the philosophical motivation for a restriction to first-order formalizations is unclear. Alternatively, syntactic theorists might be able to argue that the logical vocabulary of canonical language should include primitive interpreted set-theoretical predicates which settle such questions; a single predicate that applies only to denumerable or only to non-denumerable sets, would suffice to restore the extensional equivalence of the two views.[6]

(c) superiority of semantic conception

(i) in isolating the observational content of a theory.

That said, there are pragmatic considerations which recommend the semantic conception of theories, and the particular version of it that I have described and that is tailored to bring out the features of theories that are of particular concern in other parts of the thesis. I am not suggesting that one should wed oneself to one or the other way of conceiving theories, insofar as one has different purposes, it will be well to have several different ways of thinking of theories and to be able to switch back and forth between them, invoking whichever makes most salient features of interest in any particular context. But I think it is true that in theoretical con-

---

5. This is an immediate consequence of the Skolem-Lowenheim theorems; there is no first order theory which has models (in the logician's sense) of only one or the other cardinality.
6. Of course, if you had one, you would have the other.

texts, it is typically structural properties and relations between a theory's models that are of interest, and since these get pride of place on the semantic view, that is the one that is typically preferred. The superiority of the semantic over the syntactic conception of theories, for our purposes, derives simply from the fact that the theoretically significant properties of and relations between theories are saliently exhibited on their semantic representations, but altogether obscured by their syntactic representations.

One case in point: relations of empirical equivalence between theories. Assume (uncritically for the moment, though we will become critical in about this in the next section) that there is a distinction between what can be discerned with the naked eye and what can be seen only with the aid of instruments (microscopes, x-rays, cloud chambers, and the like). The distinction corresponds to a distinction between different substructures of the models of a theory on its semantic representation; observationally discriminable structures are represented by the empirical substructures and observationally indiscriminable structures by its higher-level structure. Two models are empirically equivalent just in case their empirical substructures (appropriately interpreted) are isomorphic, and two theories are empirically equivalent just in case their models embed the same range of empirical substructures. More precisely, a pair of theories T and T* are empirically equivalent just in case for every T-model there is a T*-model which is empirically equivalent to it, and *vice versa*. So there is a simple semantic relation the models of a theory bear to the models of an empirically equivalent theory. There is, by contrast, no such *syntactic* relation. The positivists (notoriously) thought that there was because they thought that the non-logical vocabulary of a language could be sorted into two classes: terms that refer to observable things and properties and terms that refer to unobservable things and properties. And they thought that the former was comprised of a stock of theory-neutral terms (perhaps borrowed from ordinary language) common to all of the languages in which various theories were formalized and the non-observational or 'theoretical' vocabulary was introduced in the course of theorizing and specific to particular theories.[7] It is a commonplace, nowadays, that this is a mistake; a theory can be—and typically is— formulated syntactically using terms that refer only to its basic entities and fundamental quantities, and these are theoretical terms, one and all.

Observational terms are logical constructions of non-observational terms or, perhaps, simple names introduced by definition in terms of such logical constructions, but not a special theory-free vocabulary shared by all theoretical languages.[8] It is a consequence of this that the positivists were mistaken in their syntactic criterion for empirical equivalence, for it is a consequence of this that the theorems describing the observational content of empirically equivalent theories will not in general be syntactically alike. This is really a special case of a more general fact, *viz.* that the theoretically important relations between theories are semantic relations, relations of structural similarity between the models of theories consisting in the sharing of certain substructures. Such semantic relations typically have no simple or salient syntactic representation in the languages in which theories are actually formulated.[9]

A second case in point: the pictures of theorizing suggested by the syntactic and semantic conceptions, are respectively fruitful and misleading. The syntactic conception encourages thinking about scientific theories as stories about the world; descriptions of its past, present, and future. Such structure as they have is narrative structure. We describe what we actually observe, and then fill in the holes in the story in ways that preserve its narrative structure, as though we attribute God the desire to make the world's

---

7. That we refer to the distinction as the 'observational/theoretical distinction' is an artifact of this aspect of the positivists' view.
8. As will become apparent in the next section, I do *not* want to deny that there is a distinction between the observational and non-observational vocabulary. If there is a distinction between what can be seen with the naked eye and what cannot, and it is represented semantically as a distinction between the empirical substructures of a theory's models and their higher-level structure, then it can easily be parlayed into a distinction between observational and non-observational vocabulary of the language in which an arbitrary theory is formulated syntactically. Here is how: give a semantic formulation of a theory, assuming for convenience that its models comprise an elementary class, pick out those substructures of the models which represent observable features of the world, and then determine which terms in the theory's language refer to the elements and properties which define these substructures. These comprise the theory's observational vocabulary.

    The point is rather that this vocabulary will not in general be shared by different theories, and consequently theorems of different theories which describe the same observational content will not in general be syntactically similar.
9. We can, of course, construct a language in which the features in question, once ascertained, would have a nice syntactic representation, in just the way once we have pinpointed observable objects and properties of a theory, we can construct a language in which their names are specially tagged (perhaps we let all and only names for observable things end in 'o'). This is not, however, any help in ascertaining the relevant features in the first place.

story a good one, together with the same standards of story-goodness as we employ. The narrative produced by describing experience alone has holes of two sorts, created by events that are observable but go unobserved, either because they occur too far in the past or too far away for observers to be present and convey us reliable records, or because they simply happen to go unwitnessed like the proverbial tree in the forest. In general, unobserved observable events are ones that we aren't spatiotemporally positioned to observe, but would have observed had we been appropriately placed.[10] And then there is the great big hole corresponding to the description of how things are organized 'behind the scenes', the stuff we can't observe directly, no matter how well-placed.

Theories that are empirically adequate with respect to all evidence we possess at a time t agree on the description of all observed events up until t, possibly disagreeing on how the holes in experience are filled in (observable pre-t events that went unobserved, and observable post-t events), as well as on what goes on behind the scenes. Theories that are completely empirically equivalent agree on how the holes in experience get filled in, but still possibly disagree about what goes on behind the scenes; and theories that are equivalent *tout court* agree on the latter as well, i.e. they tell the same story from beginning to end.[11] The conception of theorizing suggested by the syntactic representation, is that we describe experienced events in pure experiential terms, add sentences describing experienceable but unexperienced events also in these terms, and then add sentences in quite different terms 'explaining' experience with hypotheses about what gives rise to it.

---

10. More precisely (if precision is to be had by such means) s is observable iff there is some spatiotemporal position **p**, such that if we occupied **p** and were attentive, we would observe **s**.

    Assessing the truth value of the counterfactual is dicy because the antecedent may express a physical impossibility (e.g., if s exists outside our event horizon, or in the future, or if it is so far away it takes more energy than there is in the universe to get there...), but what we should do is hold fixed everything we can while making the antecedent true, including our abilities and the laws governing our sensory interactions, but breaking - in a local way - whatever laws are necessary to put us at p.

11. This isn't quite right, because a physical theory won't in general tell a single story about the way the world is, but instead provide a family of related stories which depict ways it *could have* been, i.e. stories which describe physically possible worlds, one among which is the actual.

The semantic conception suggests a quite different conception of the relation between what is revealed in experience and the underlying structure depicted by our theories, between the manifest image and the scientific image, in Sellars' evocative phrase. Instead of the emphasis on *entities* too small to see, it suggests that we should think in terms of structures that cannot be discriminated with our bare eyes. We should investigate the conditions under which it is appropriate to discern hidden structure, differences between systems which appear alike to the naked eye. Instead of conceiving the process of theorizing as one of adding to the chaotic happenings on-stage a description of the proceedings behind the scenes which give rise to them, we should conceive of it as analogous to painting a picture of a scene viewed from afar, and the differences between empirically equivalent theories as differences in the fine-scale structure of the depicted which have been improvised by the artist. And instead of thinking of the observational/theoretical distinction as applying to entities, and theoretical entities as entities which happen to be too small to see, we should think of it as applying to properties, and theoretical properties as those which mark intrinsic differences between systems which appear perfectly similar. Embedding the appearances in models with higher-level structure, we are not postulating the existence of invisible entities, but recognizing intrinsic differences between systems which appear similar, but behave differently.

The syntactic view encourages one to think of theories as stories, the evidence for theories as a straightforward transcription from experience, and scientific theorizing as filling the gaps in the spotty and episodic narrative provided by experience. On the semantic view, by contrast, experience is regarded as providing a glimpse of the coarse-grained structure of the world, hints at the fine-grained structure depicted in our models. We start with an empty canvas and brush—in broad strokes—the outlines of what is actually seen. Next, we guess at the outlines that fall outside our field of vision at the same crude level of representation, and brush them in with strokes as broad. Finally, we go back and fill in the small-scale structure. Theories which are theoretically distinct but empirically equivalent, then, are not so much like stories which agree only in parts, but like pictures which agree on the rough shape of a depicted scene but disagree over the detailed structure.

To what extent do these associated 'pictures' actually guide our philosophical thinking about science? and what effects do they have? These are subtle questions and hard to gauge, but important for assessing the relative merits of the two conceptions. A look at the history of the philosophy of science will, I think, be persuasive that a great many of its mistaken ideas were shaped by the way of thinking encouraged by the syntactic conception of theories.

(iii) in understanding methodological criteria for theory choice.

A third case in point: the non-empirical methodological criteria at play in theory choice, is not a syntactic one. Go back to the stage of history in which the only bit of evidence we have which favors Einstein's theory of gravitation over Newton's is the advance of the perihelion of mercury. Consider a syntactic formulation of a theory which is—word for word—like Newton's save that it contains an additional sentence which states the exception needed to accommodate Mercury's orbit and restore the theory to empirical adequacy. What favors Einstein's theory over this revised Newtonian theory? (They make different predictions but we are supposing that these are untested and asking where one should place one's bets at this stage). Although we would have been hard put to choose between Einstein's theory and Newton's own, and although the revised Newtonian theory is not significantly different *syntactically* from the original and so does not measure up significantly differently against syntactic criteria of simplicity, elegance, etc., the revised Newtonian theory loses out by a long shot against Einstein's.[12] Whatever methodological criteria we have for choosing between theories, they do not clearly favor Einstein's over Newton's original theory, but they do clearly and obviously favor Einstein's theory over the revised Newtonian one. Since the original and revised Newtonian theories are so syntactically similar, this suggests that the criteria—whatever they are—are not syntactic.[13] One wants to say that the revised Newtonian theory is ad-hoc and Einstein's theory is not, but this does little

---

12. Feynman remarks *a propos* of Newtonian gravitation theory and the perihelion of Mercury:

> "Newton's ideas about space and time agreed with experiment very well, but in order to get the correct motion of the orbit of Mercury, which was a tiny, tiny difference, *the difference in the character of the theory needed was enormous.* ... In order to get something that would produce a slightly different result it had to be completely different." (*The Character of Physical Law*, p. 71)

more than name the feature of the latter which recommends it over the former. The real question is whether we can characterize the feature in syntactic terms, i.e. whether we can say purely on the basis of their logical form what makes one sentence an *ad-hoc* addition to a theory and another a well-motivated revision. This seems hopeless; for given any sentence S which expresses a well-motivated addition to a theory T, it is easy to find a sentence S', with the same logical form as S, which would be an obviously *ad-hoc* addition to T.

In general, wherever we have a very few easily describable experimental results which conflict with an otherwise well-supported theory, we can produce a theory which accommodates the phenomena by adding to the original theory a short and simple sentence describing the exceptions at no great cost to syntactic simplicity, elegance, etc., and which produces a theory that is very syntactically very similar to the original. What is wrong with such theories is not that they do significantly worse than the original by any standard that deserves the name of syntactic simplicity or elegance, or that is even easily characterizable in syntactic terms. What is wrong with such theories is that they do much worse by a peculiarly semantic criterion I try to spell out in section 5.6. It is nothing like our intuitive notions of syntactic simplicity, nor any of the more precise proposals for capturing the syntactic notion (except insofar as these are tailored to make sense of scientific judgments and trade tacitly on the semantic criteria), though I think it does have some claim to being called semantic simplicity, simplicity of structure of a theory's models. More on this in section 5.6.

To recap: the syntactic and semantic conceptions of scientific theories are almost formally equivalent: given the set of a theory's models-in-the-sense-of-the-semantic-view, we can arrive at a syn-

---

The 'enormous difference' in the 'character' of Newton's theory which he says is required to accommodate Mercury's orbit, is not evidently a syntactic one, for we have seen that a very small syntactic change which will do the trick. What Feynman has in mind here as the 'character' of a theory is evidently not revealed in its syntactic formulation, for enormous differences in character correspond to relatively minor differences in syntactic formulation.

13. Nor is this simply an instance of the problem of old evidence (the problem posed by the fact that evidence provides support for a theory only if it was predicted by the theory before it was gathered); Einstein's theory would be better than the revised Newtonian one even if the latter had been proposed before, and the former had been proposed after, the discovery of the perihelion of Mercury.

tactic formulation in any chosen language. Likewise, we can recover the models-in-the-sense-of-the-semantic-view of (almost) any theory from a syntactic formulation of it. The exception occurs only when the theory in question has only denumerable or only non-denumerable models-in-the-sense-of-the-semantic-view (i.e. when it says that there are only denumerably many or only non-denumerably many things), for there is no way to express this fact syntactically in a first-order formulation. This difference is not, however, decisive between the two views. The real superiority of the semantic conception lies in the fact that theoretically important properties of and relations between theories (typically, the sharing of various substructures) are saliently exhibited on their semantic representations, but altogether obscured by their syntactic representations. I mentioned three important cases:

(i) the relation of empirical equivalence between theories (sameness with respect to observational content) has a simple semantic representation, and no simple syntactic representation,
(ii) the pictures of theorizing suggested by the syntactic and semantic conceptions, are respectively fruitful and misleading, and
(iii) the non-empirical methodological criteria we employ in theory choice are not syntactic but semantic.

For most of what follows, I'll speak in terms of the semantic view, sometimes translating into terms that apply to theories as conceived on the syntactic view where it is convenient and where the syntactic formulation is more familiar.

3. OBSERVABLE/UNOBSERVABLE DISTINCTION

(a) making out the distinction

There can be little doubt that there is a distinction between what can be seen by unimplemented sight and what can't, and the distinction cannot be argued away. Where precisely the line between the two lies is an empirical question that can be determined by more or less subtle tests and that depends, in part, on the sensitivity of our natural sensory apparatus. It has, however, sometimes been asked to bear a philosophical weight that has made it the object of some very powerful attacks. The positivists, for example, held that it marked a fundamental divide between physical truths that we can grasp and those that are beyond our ken. More recently, Van Fraassen has argued that, while we can *grasp* hypotheses about what cannot be observed directly, we can never have evidence which favors one over another, so there is no rational route to opinions about what cannot be directly observed.[14]

Traditionally, the distinction is taken

(i) to apply to entities,
(ii) to be a distinction between macroscopic and microscopic entities; things we can see with our bare eyes and those which need to be magnified (tables and chairs, as opposed to cells, electrons, and the like), and
(iii) to coincide with the distinction between the denotations of a neutral vocabulary common to all scientific theories and the denotations of terms tied to a specific theory and introduced in the course of theorizing.

There are several things now widely recognized as wrong with the traditional way of drawing the distinction. First, (iii) is false for reasons that I mentioned in the last section. Scientific theories depict the world as comprised of a distribution of quantities over a set of basic individuals. The basic individuals are typically microscopic in size, and the quantities bear little resemblance to the kinds of observable qualities which characterize ordinary objects. These latter are identified with particular constructions out of basic quantities, and macroscopic objects are identified with large clusters of more basic individuals. One most naturally associates a theory with the first order language whose non-logical vocabulary consists only of predicates referring to the values of the basic quantities. In such a language, the distinction between observational and non-observational vocabulary will obviously not coincide with a distinction between vocabulary which is theory-neutral and that which is theory-laden, for there will not typically *be* any theory-neutral vocabulary. Terms which refer to observable things will be

---

14. At least none that is rationally *compelled*; van Fraassen is adamant about the distinction.

more or less complicated constructions out of the theory-laden (and theory-specific) terms referring to the basic quantities. Nor will there be any vocabulary which is common to all theories. Insofar as the choice of basic quantities is theory-specific, the non-logical vocabulary of a theory's language will be so as well; the language of distinct theories may have no terms (save logical terms) in common at all.

The problem is that, while we want to say that the observed facts are epistemically more basic, they are—by the lights of one or another—ontologically less so. The observable facts are what you get when you describe the world at a certain level of abstraction, from a certain distance, but the language of a theory is built to describe the world from the bottom up. Since it is what the world looks like at the bottom level over which theories disagree, the non-logical vocabulary of the languages in which they are formulated will typically be disjoint, no matter how well they agree at other levels of description.

I said one 'naturally associates' a theory with such a language, though of course a theory can be formulated in any language that is sufficiently rich. One might try, in light of this, to argue as follows: everyday language contains terms which describe the ordinary observable world, in fact, we would expect it to contain a rich enough stock of such terms to characterize it in complete observable detail. And we would expect these terms—given that they've survived innumerable changes in theory—to be relatively insensitive to shifting theoretical opinions about the physical constitution of those things. If this is right, then the fragment of everyday language containing these should comprise a neutral language in which we can state the evidence for our various theories and compare them one against the other. We should, that is to say, use this fragment of natural language as neutral terms in which to present a theory by defining the theory-specific vocabulary in terms of it (implicitly or explicitly), and then stating the theory's laws in the defined vocabulary.

The problem with the suggestion is that ordinary language is far from a single, well-defined monolith shared by all speakers. Nor is the part of English which is used to talk about the observable world theory-free. Think of all of the ordinary terms referring to observable things that have their roots in scientific theories; 'radio', 'micro-wave', 'anti-biotic', and so on. Feyerabend puts the point nicely:

> It has been frequently asserted that the language in which we describe our surroundings, chairs, tables and also the ultimate results of experiment (pointer-readings) is fairly insensitive towards changes in the theoretical 'superstructure'. It seems somewhat doubtful whether even this modest thesis can be defended; first, because a uniform 'everyday language' does not exist. The language used by 'everyday man' is a mixture of languages, i.e. a means of communication which has received its interpretation from various and often incompatible and obsolete theories. Secondly, its is not correct that this mixture does not undergo important changes: terms which at some time were regarded as observational elements of 'everyday language' (e.g., 'devil') are no longer regarded as such. Other terms, such as 'potential', 'velocity', etc., have been included in the observational part of everyday language, and many terms have assumed a new use. ("An Attempt at a Realistic Interpretation of Experience", p. 30)

Even ignoring these points, i.e. even assuming that there is a part of our (more or less) common vocabulary which is used to talk about the observable world and agreeing to ignore whatever theoretical associations it has, it is hard to see what purpose is served by requiring that it be incorporated into each of our scientific languages (all the theoretical terms referring to observable things identified with some term in the common vocabulary, and every theorem involving only observable things equivalent to one containing only common vocabulary). It couldn't provide us with a workable syntactic criterion for isolating the observational content of an arbitrary theory, for before we could implement it, we would need (i) an independent means of identifying the observational vocabulary in English, and (ii) definitions of each of the terms in the common vocabulary with theoretical terms in the language of the theory. There is no reason to think that the former is any easier than identifying the observational vocabulary in the theoretical languages directly. In fact, we would need to do as much, as well as to know what each refers to, in order to provide (ii). If there were such a fragment of natural language, it *would* give us a common language in which to compare the observational content of different theories, but given the existence of alternative, relatively language-independent ways of representing and comparing content, I don't see what advantage it harbors. It is, in short, a difficult and imprecise way of doing something (identifying and comparing the empirical content of different theories), that can be done more simply and precisely otherwise.[15]

Any distinction between observational and non-observational vocabulary is derivative of a distinction between the objects (entities, properties, and relations, alike) to which they refer, so we do better to simply concentrate on the objects themselves. It is unexceptionable that there are entities we can see and entities we cannot, and the notion that the difference between them is an epistemologically significant one has an immediate and intuitive appeal. Surely, our knowledge of the familiar things we bump up against, the ones we touch, see, and smell, the tables we use, the cars we drive, and the people we know, is ever so much more secure—indeed fundamentally different from—some of the entities postulated by science, of which we see only the most ephemeral traces and then only under the most contrived circumstances: tracks in a cloud chamber, dots on a photographic plate. We are 'in direct contact with' the former, but our knowledge of the latter is derivative and indirect.

The objections to the idea are so familiar nowadays as to seem hackneyed, and I won't recite them. There are a number of different ways of bringing it out, but the basic point is that observability abstracts from spatiotemporal limitation on observation, but there is no reason why sensory limitations should have a different status. Just as changing our spatial location brings some unseen things into sight, augmenting our vision with instruments brings other unseen things into sight, and there is no obvious reason that our knowledge of the latter should be less secure than our knowledge of the former. Churchland makes the point nicely by imagining a race of creatures who are physiologically just like us except that, as it happens, they are naturally endowed with the kinds of instruments *we* have to *construct* to augment our senses.[16] Specifically, they are born with an electron microscope implanted over their left eye; when they close it, they see as we do, when they close their right eye, they see what we see through an electron microscope, and they say the same things that we do about the world, both its observable and unobservable parts. Surely, the reasoning goes, such creatures are in no different epistemological position than we; the difference in the origin of our respective viewing instruments—natural in their case, manufac-

---

15. In terms of the distinction between the empirical substructures of a theory's models and their higher-level structure.
16. "The Ontological Status of Observables" in *Images of Science*, eds. Hooker and Churchland, p. 35).

tured in ours—doesn't make for a difference in their epistemological relation to the entities they are used to examine.

There is a whole literature in which these sorts of arguments are elaborated and I think them entirely persuasive, so far as they go, but that they miss something because they take it for granted that the distinction between the observable and the unobservable is properly characterized by (i) and (ii). It is natural to focus on entities, and to think that the difference between what can be observed and what can't is a difference in size, specifically, that it is only things that are too *small* that cannot be seen, but the focus is misplaced. First, because it is not only the over-small that is unobservable, but the over-large as well. The whole Pacific Ocean is just as outside our observational ken as a single molecule of H_2O. Just as we typically detect properties of microscopic entities indirectly by seeing properties of the collections of them that correspond to macroscopic objects, so we usually detect properties of very large-scale phenomena by seeing—piecemeal—the properties of their parts.[17]

Second, very little is really too small for us to see. Humans can apparently see light from individual photons; place us in a dark room, allow our eyes are allowed to dark adapt, expose us to very weak flashes of light, and ask us to indicate when we see flashes. When the statistics of our detections is compared with the 'shot noise' of the photons, it turns out that we are remarkably good detectors of the presence of even just a few. It is true that we are not perfectly accurate, but the accepted theoretical explanation of our mistakes is as remarkable as our sometimes success, for it is thought that some of the false positives are caused by thermal fluctuations exciting the rhodopsin molecule which is usually only excited by light![18] If this is right, it means that we are actually perceptually sensitive, under the right conditions, to thermal fluctuations.

Third, not *all* properties of even medium-sized things, are observable; we don't, for example, see the gravitational or magnetic properties of ordinary objects. Indeed, our senses are

---

17. I sometimes use 'see' where I mean more generally 'perceive'.
18. The evidence for this comes from experiments done with frogs. Unlike people, frogs are cold-blooded, so their body temperature is the same as the ambient room temperature, and the temperature of the frog's retina can be controlled by varying the temperature of the room in which they are situated. If we perform similar experiments with frogs, training them to stick out their tongues when they detect a flash and exposing them to weak flashes, we would expect, if indeed a significant proportion of their misdetections were due to excitations from thermal fluctuations, that there would be an exponential decrease in the number of observed flashes in the absence of light as their retinal temperature is decreased. This is in fact what is seen.

'attuned' to only a very small and select set of properties of such things, and the real difference, the distinction that matters for the purpose of separating the observational from the non-observational content of a theory, is the distinction between properties to which we are sensitive and those to which we are not. I will have more to say about this later, but first I'll see if it is possible to do a better job of making out the distinction between the observable and unobservable. The first thing we have to do to make out the distinction is recognize that it applies directly to properties, and only derivatively to entities. The second thing is to see that it applies not to individual properties, but to *distinctions between* those in a family whose members are mutually exclusive and jointly exhaustive, i.e. a family of properties whose members have the internal structure of the cells in a partition. The observable/unobservable distinction applies in the first instance to quantities, and only derivatively to properties and entities. A quantity is observable just in case we can distinguish its values from one another by sight, i.e. *iff* unimplemented observation, under the right conditions, (in full daylight, at a close distance, with full attention, ... the only constraint on conditions is that they not include sensory implementation) suffices to determine which of its values obtain.[19] Simply put, a quantity is observable just in case different values *look different* to ordinary observers in ordinary conditions (where, again, ordinary just excludes sensory implementation).[20] The values of an observable quantity are the particular types of smells, tastes, textures in terms of which the world presents itself to the

---

19. The basic idea here, with a slightly different emphasis, is the one behind Feyerabend's pragmatic theory of observation. The idea that it is *distinctions between* situations to which the notion of observability applies, and not individual situations, is his also. He writes:

> "Whether or not a situation s is observable for an organism O can be ascertained by investigating the behavior of O,...it can be ascertained by investigating O's ability to distinguish between s and other situations. And we shall say that O is able to distinguish between s and situations different from s if it can be conditioned such that it (conditionally or unconditionally) produces a specific reaction r whenever s is present, and does not produce r when s is absent...." (Feyerabend: "An Attempt at a Realistic Interpretation of Experience", p. 19).

Feyerabend's emphasis is metalinguistic (he is mostly concerned with isolating the observational terms of a language and making some points about its Interpretation), and he speaks in terms of situations rather than quantities (where situations are attributions of properties to objects), but these are superficial differences.

senses. We usually think of these as qualities, and I will henceforth refer to them as such, but I want to dissociate the term from the mentalistic associations that usually accompany it. Qualities, in the sense that I am using the term, are physical properties, ordinary properties of physical things, or—if you like—sets of physical objects. Color is an observable quantity because different colors look different to ordinary observers; refractive index or charge is not. Goodman's gruller is not an observable quantity because grue things observed before Jan. 1, 2000 and bleen things observed thereafter do not look different to ordinary observers.[21]

(b) observation as measurement

The distinction between observable and unobservable quantities, then, is a distinction between those quantities whose values can be distinguished by unimplemented observation and those whose values cannot, i.e. between those whose differences in values correspond to qualitative differences and those which do not. Let me

---

20. Consider a quantity q, and suppose we want to know whether q is observable by the above criteria. Recall that we are assuming that we have a fully formulated theory and trying to say what distinguishes the observable quantities from the unobservable ones, not trying to show how to come up with a list of basic quantities given only a set of objects and various kinds of observational distinctions between them, so we can assume that we know what the full set of basic quantities and the values any object takes for them, and we can restrict our attention to partitions of objects which are perfectly similar save for their q-values.

   Objects falling in the same cell of the q-partition can be observationally distinguishable (this is will the case if there is no more fine-grained version of q which is an observable quantity), but need not be (any coarse-graining of an observable quantity is also observable, and the observationally distinguishable objects will in general share cells of any coarse-graining of an observable quantity).

   Must we be able to distinguish observationally *any* two objects in different cells of the partition defined by q for it to count as observable, or must it just be the case that there are *some* observationally distinguishable objects in different cells? The latter, I think. The objects in the individual cells - whether or not they are observationally distinguishable - will cluster around central cases, and q is observable just in case the central cases in different cells are observationally distinguishable. Think of the case of color: shades immediately adjacent to one another on the color spectrum are not observationally distinguishable, but there are coarse grainings of quantities defined by precise coordinates in the color spectrum that are observable. These are just the quantities which define partitions in which the central cases in different cells are observationally discriminable. (the color spectrum in this case, provides the measure of closeness between colors that gives precise meaning to the notion of a 'central case'.

21. Goodman's definitions:

    grue=$_{def}$ green if being examined before Jan. 1, 2000, blue if being examined afterwards.

    bleen=$_{def}$ blue if being examined before Jan. 1, 2000, green if being examined afterwards.

put this in a slightly obscure way, which I think is nevertheless illuminating for reasons that will become clear later. Think of us as mobile natural systems whose sensory apparatus' act as measuring instruments for the observable quantities in our immediate environments. What makes a quantity observable is that our natural sensory apparatus 'attuned to them' in the sense that we end up in different states, states that we can discriminate from the inside, depending on which value of the quantity in question obtains.[22] Relations of similarity and difference between states of being-thus-and-so-appeared-to are taken as primitive, our judgments about them are epistemologically basic in the sense that they are not inferred from anything else. Qualitative differences between objects are differences in how they appear; objects are qualitatively different just in case there are ordinary conditions under which interacting with them puts us in states of being-differently-appeared-to.

It is a happy legacy of the interpretive difficulties of quantum mechanics that there is a large body of philosophical literature devoted to the notion of measurement. The discussion is sometimes technical and there are some difficult questions about details, but there is broad agreement on this much: a system Y is a measuring apparatus for quantity **A** *iff* Y has a certain possible state (its ground-state) such that if Y is in that state and coupled with another system X in any of its possible states, the evolution of the combined system (X + Y) is subject to a law of interaction which has the effect of correlating the values of A in X with distinct values of a quantity **B** in Y. The Y-quantity **B** is called the 'pointer observable', and its values indicate X's **A**-value. We can define observability in these terms as follows: a family of properties $\mathbf{A} = \{\mathbf{A}_1, ... \mathbf{A}_n\}$ is **observable** *iff* different $\mathbf{A}_i$'s are distinguishable (to ordinary people under ordinary conditions) by sight; **A** is observable, that is to say, iff unimplemented observation counts as a measurement of **A**. We can also say in these terms, what it is for an unobservable quantity to be nevertheless measurable; a quantity B is measurable iff there is some system which can act as a measuring device for **B**, the pointer observable of which is an observ-

---

22. I won't worry too much about what these different states *are*, i.e. whether they are merely different dispositions to answer one way or another if queried about which value of the quantity obtains (they are at least that), or different mental states, or - perhaps - different brain states.

able quantity. So **B** is measurable if unimplemented observation counts as a measurement of the pointer observable on a system which, in its turn, measures **B**.

From a physical perspective, the picture is this: observations are divided into types according as their objects appear different (to ordinary observers, under ordinary conditions), or—as I put it, to emphasize the analogy with measurement—as we end up in states of being differently-appeared-to upon interaction with them. An observable quantity is one distinct values of which appear different; an *un*observable quantity which is nevertheless measurable is one whose values are tracked by the values of some observable quantity, typically under very special controlled conditions, *viz.*, the fixed conditions inside the measuring instrument of which the observable quantity in question is the pointer observable. Whenever there exists a physical process which correlates different values of an observable quantity with different values of an observable one, even if the conditions are highly contrived, and even if the correlation is short-lived, so long as it lasts long enough to permit an observation of the latter, the unobservable quantity in question can be measured, for in such situations the value of the observable quantity indicates that of the unobservable one. This is the essence of measurement; the idea behind it is to exploit nomological relations between observable and unobservable quantities to transform the latter into observable form. The transformation can be as attenuated and technologically complex as you please, the only requirement is that it correlate the different values of the measured quantity with observationally distinguishable effects under reproducible conditions. Creating the conditions under which the effect is produced, then, can function as a test for the presence of the property in question.

The basic quantities of our physical theories are typically unobservable, but measurable. There is a third class of properties, 'unascertainable quantities', that are neither observable nor causally integrated with observable quantities in a way that permits measurement. These have no discernible effect on experience; experience (even combined with complete knowledge of the physical laws) can tell us nothing about the distribution of any unascertainable quantity, nor can the distribution of any observable quantity tell us anything about experience. For this reason, unascertainable properties have little part in science, and I will have correspondingly little to say about them. The distinction between ascertainable and unascertainable is the distinction between what can be determined on the basis of experience and what is forever beyond our experiential ken, and it marks a deep epistemological divide; the distinction among ascertainable properties, between those which are observable and those which can only be measured, on this way of seeing things, does not. Human beings, on this picture, are mobile natural systems whose sensory apparatus' act as measuring instruments for some of the quantities in their environments (the observable ones). We can be fitted up with special attachments that allow us to measure additional quantities in the same way that a camera or a microscope can be fitted with lenses to make it sensitive to special types of light, or properties of objects at greater distances. Observations, or appearings, sorted into primitive types (ways of being appeared to, distinguished phenomenally) track observable properties, and supplementation with microscopes and other imaging instruments widens the range of quantities whose values they track. Our relation to these additional quantities is no different in principle than our relation to those we detect with the unaided senses; the former require more sensitive instruments than those with which we are by naturally endowed, but we know how to overcome our natural limitations.

A crucial point in all of this is that—whatever we think of the relation between observations and the quantities they are observations *of*—on some level we have to appeal to a primitive division into kinds. Even if we regard observation as a sort of measurement and the relation between observations and observable quantities as akin to that between the positions of a pointer observable and the values of the quantity it indicates, it should be obvious that we don't get off the ground until we have divided the values of the pointer observable into kinds. I will have more to say about properties in general and about these primitive kinds, in particular, in a later section, but now I'll move on.

(c) imaging instruments

I've said that if we think of perception as a physical process, then perceptual states (states, descriminable from the inside, of being thus and so appeared to) relate to observable properties as the different values of a pointer observable relate to different values of a measured quantity; perceptual states, that is to say, indicate the values of observable quantities, and observable quantities, in their turn, act as pointer observables for other measurable quantities. The relations of covariance (between perceptual states and observable quantities, and between observable quantities and measurable ones) which make all of these possible, are consequences of the physical laws, but it is more complicated than it might at first sound. Even in the deterministic case, where **A** is a measured quantity and **B** the pointer observable which indicates its value, we never have a simple equational linkage of the form **B**=f(**A**). More typically, **B** is a function of a set of quantities which include **A**, and the trick of measuring **A** is to set up a physical situation in which we hold the values of other relevant parameters fixed, so that variations in **B** can be traced to differences in **A**. Only in these special controlled circumstances, do **B**-values track A-values. It would be more accurate to think of the pointer observable as the complex quantity, **B**-in-these-special-controlled-circumstances or **B**-with-such-and-such-values-of-quantities-$Q_1...Q_n$. This means thinking of the pointer observable not as simply the position of the dial on the measuring apparatus, as is customary, but rather as the position of the dial on the measuring apparatus under conditions which include the specification of values of other quantities, and there may be an indefinitely large number of these.[23]

Be that as it may, if this is the right way of thinking, there is not an epistemologically interesting difference between unimplemented sight and sight augmented by imaging instruments.[24] What we take away from either type of interaction is information about the *structure* of an object with respect to some quantity which is iden-

---

23. The indeterministic case is not much different, if we count the chances, or expectation values, as observables pertaining to large ensembles of systems (bearing some logical connection with frequencies in the ensemble, of events that they are chances *of*), and agree to reserve the term 'measurement' for interactions that give information about the state of the system on which they are performed. In particular, we should agree not to apply the term to the individual interactions we ordinarily think of as measurements (e.g. sending a single electron through a Stern-Gerlach magnet, or an individual photon through a filter), and insist that only a large enough set of these individual interactions counts as a measurement, different values of which distinguish *ensembles* in which frequencies (or expected frequencies as the size of the ensemble approaches infinity) differ. There is good reason to deny that the individual interactions are measurements in any sense, for they yield almost no information about the systems they are supposed to be measurements on (they yield only the information that there is a measure zero probability that the system was, before measurement, in an eigenstate of the measured observable, orthogonal to the observed value).

tified by its nomological relations to types of appearings. The same is much more obviously true of ordinary measurements where the output of the instrument is not an image but a more abstract representation like a set of numbers or a graph. The psychological difference between a process which presents us with a picture—in living and breathing color, so to speak—and one which presents us with a bare mathematical description, is *just* a psychological difference; both contain the same sort of information. Seeing, whether with our bare eyes or through a microscope, is just measuring; what distinguishes it from other types of measurement is only the pictorial form in which its results are presented.

It might help to make the discussion a bit less abstract, and to get a sense of the relation between their objects and the images they produce, if we say something about how vision and some different types of microscopes actually work. In the case of ordinary vision, it is usually the light reflected by an object that hits our eyes and produces a retinal image. Sometimes, however, e.g., if we are using a magnifying glass to look at a specimen illuminated from behind, the image is formed instead by transmitted or absorbed light. In either case, patches of dark and light in the visual image correspond to the proportions of light transmitted or absorbed by parts of the object under view, so what the naked eye detects are changes in the amplitude of light rays due to differences in the reflective properties of matter. A simple light microscope works a little differently; it produces an image by synthesizing rays of light *diffracted* by the specimen under the lens, so what shows up in the image are differences in the way parts of an object diffract light. Matter, especially living matter, often varies in its birefringent (or polarizing) properties, and differences in these are detected by polarizing microscopes.[25] Likewise, certain biologically important substances have characteristic ultraviolet absorptions which are revealed by ultraviolet microscopes, and phase contrast microscopes reveal differences in *refractive* index of various parts of a specimen, by converting them into visible differences of intensity in the image.[26] The interference contrast microscope also makes

---

24. Or, better, imaging instruments coupled with procedures which prepare the specimen by augmenting it with dyes and such which bring out the properties of interest.
25. The polarizer transmits to the specimen only light of a particular polarization and, in the simplest case, the analyzer is placed at right angles to the polarizer. If the specimen is birefringent, it changes the plane of polarization of the incident light and a visible image is formed.

differences in refractive indices visible, but works quite differently and—unlike the phase contrast microscope—provides a quantitative measure of refractive index.[27] All of these techniques (and variations produced by combining them, e.g., polarizing interference microscopes, multiple beam interference, phase modulated interference, etc.) exploit the sensitivity of light waves to various different *quantities* of matter; they use the fact that a light wave interacting with a specimen *bears the marks* of the spatial distribution of the quantity over the specimen, in the sense that the values of quantities pertaining to the wave are correlated with the relevant specimen quantities after interaction.

Sound waves also interact in interesting ways with certain properties of objects, and can be used in a similar fashion to construct acoustic microscopes. Radar, for example employs longitudinal wave-fronts to detect objects outside the visual field of airplane pilots; sonar uses transverse wave fronts to the same effect for naval captains. Ultrasound works by converting sound signals of very high frequency which have interacted with a specimen first into electrical signals and then into a spatial display on a screen. Unlike light, sound waves can be transmitted through opaque objects, and hence can be used in *in vitro* examinations of fetuses, in metallurgy, and the like. They are also sensitive to density, viscosity, and flexibility of living matter, and—since they don't damage a specimen—can be used to monitor changes in these during processes *live*. Besides acoustic and light microscopes, there are also, of course, electron microscopes, at least as many different kinds of these as of the former, and others. All work in the same way, by exploiting the fact that some quantity pertaining to a wave of one or another sort is sensitive to differences in a quantity pertaining to a specimen, to produce an observable image of the specimen which discloses the spatial distribution over it of the quantity in question. A little more simply, and in terms recognizable from above, since the value of quantity **A** pertaining to a wave is,

---

26. Whereas in an ordinary microscope the image is synthesized from the diffracted waves D and the directly transmitted waves U; in the phase contrast microscope, these are separated and one is subjected to a phase delay which produces in focus phase contrasts corresponding to the differences in refractive index in the specimen.

27. They function as follows: a half-silvered mirror is used to split the light source; half is sent through the specimen and half is kept as a reference wave to be recombined for the output image. Changes in optical path due to different refractive indices within the specimen produce interference effects with the reference beam.

after interaction with the specimen, correlated with the spatial distribution of quantity **B** over the specimen, the wave can be pressed into service to produce an image of the specimen which reveals its spatially varying **B**-values. *U*nobservable quantities, quantities which do not make for differences in transmitted or reflected light of a sort to which our bare eyes are sensitive are in this way *made visible* by converting them into differences which *do*.

The image produced can be a simple scaled version of the specimen, or it can bear to it some more complicated relation (electron diffraction microscopes, for instance, can produce images in reciprocal space [conventional space 'turned inside out', so that near is far and far is near], and von Laue diagrams of crystals produced by x-rays aren't anything like simple pictorial representations of their molecular structure, though they can be used to generate such pictures. An image is a kind of map; it encodes information about the geometric structure of an object in its own geometric structure (parts of the image are mapped onto parts of the object, and the spatial configuration of the parts of the object are reflected by that of the corresponding parts in the image). The same information can be encoded in other ways (e.g., by a set of numbers or a set of sentences), and any process whose output, when 'fed' a specimen, encodes the relevant information, in whatever form, can be turned into one that produces a proper *image* of the specimen, by tacking a computer (or lab assistant) onto the end which translates the information into a proper little picture. The relation that the image one sees through the lens of a microscope bears to the specimen is this: the spatial configuration of the image's observable properties provides a map of the distribution over the specimen of the unobservable quantity which is revealed in the process.

Allow me to pause to make a point which is most simply made here, but whose relevance will only become clear later. There is a set of objections to views which recognize a deep and principled difference between our knowledge of the unobservable and our knowledge of the observable features of the world. They are relevant here because I think they're revealing in a different way than that intended by their authors, one that lends some support to the application of the observable/unobservable distinction directly to quantities rather than entities. Grover Maxwell made the point early on against the positivists, that our relations to tables and electrons, respectively, lie at two ends of a spectrum beginning with what can be seen without any sort of visual augmentation and ending in what can be seen only with the help of the most technologically sophisticated instruments. When we focus on cases which lie close to each other at any point along the spectrum we are hard put to convince ourselves that there is a deep epistemological divide between them. As he put it:

> there is, in principle, a continuous series beginning with looking through a vacuum and containing these as members: looking through a windowpane, looking through glasses, looking through binoculars, looking through a low-power microscope, looking through a high-power microscope, etc., in the order given. (G. Maxwell, "The Ontological Status of Theoretical Entities, *Minnesota Studies in Philosophy of Science*, 1962, p. 7)

The upshot is supposed to be that the division between what can and what can't be seen by unimplemented sight is—as Seager put it in a recent article—"the realist's beachhead, from which an assault deeper into the territory of the unobservable would be hard to contain", but that it is just not a psychologically plausible option to think we bear a fundamentally different kind of relationship to objects which lie on other sides of but close to the line between them.[28] The point can be pressed with examples; here is one from Seagar:[29]

> A simple and beautiful [pair of images] can be found in [a biology textbook by Holldobler and Wilson, 1990. p. 231], in which we see the hind end of an ant at two magnifications, revealing the details of several pheromone receptacles. The first image is not in fact utterly microscopic, but it reveals detail that certainly is beyond the range of human vision. The second image then shows details of some pheromone receptacles at ten times the magnification of the first, and these are entirely microscopic. (p. 467)

"Personally", he writes, "I find it almost impossible to resist the conviction that the receptacles are there, more or less as imaged." The Chiharas elaborate, to the same end but in more detail, an example having to do with the indiscernible mouths and legs of barely visible mites, to the same end.[30] Both are convincing so far as they go, but what these authors don't discuss is the difference between such examples and the more abstract structures of math-

---

28. The example is taken from Seager, "Ground Truth and Virtual Reality: Hacking vs. van Fraassen", *Phil. Sci.* 62, 1995, p. 468.
29. Seager, *ibid.*, p. 468.

ematical physics: electromagnetic waves, quarks, etc., which have a psychologically very different status. Putnam remarks somewhere that even a child can understand talk of people that are to small to see, and he is surely right, but a child cannot understand talk of fields and quarks, and we can legitimately wonder why. What is the relevant difference?

The answer falls quite nicely out of the account that I have given: the properties which characterize the former are observable (or some not-too-complex gerry-mandering of observable ones), whereas the latter are things with complex mathematical definitions in terms of observable properties, as complicated in most cases, as the theories in which they are embedded. A child can understand talk of people that are to small to see, because if asked what they are 'like', she can be told that they are just like us, only smaller. She cannot understand talk of fields and quarks because she does not know the mathematics needed to say what 'they are like' in terms that she can understand, i.e. in the qualitative terms that are, for all of us, the touchstone of understanding.[31] That is the difference.

4. CONTENT OF THEORIES: INTERPRETATION

(a) interpretation on semantic view; charity

In section 2, I said that I preferred the semantic view of theories according to which a theory is a set of interpreted models, i.e., a set of structures together with a mapping of elements and relations of the structures onto physical elements and relations. The empirical substructures of a theory's models are those defined by the observable quantities. These are embedded in higher-level structure of ascertainable (typically unobservable, but measurable) quantities. The empirical content of a *model* is contained in its empirical substructures; the empirical content of a *theory* is that the empirical content of one of its models accurately represents

---

30. *BJPS*, Dec. 1994.
31. Even energy is a bit of a stretch. When I asked Gideon Rosen's 3-year old daughter, Gracie, what energy was, I got the beginnings of a theory: she said 'it makes things go', and when I pressed her about what counts as 'going', she agreed that radios are 'going' when they are making noise, a fan is 'going' when the blades are turning, an oven is 'going' when it is hot …but her ideas soon gave out.

the observable structure of the actual world. Two theories are empirically equivalent just in case they embed the same range of empirical substructures, and—of course—a theory is empirically adequate with respect to the evidence we actually possess just in case one of its models (the one representing the actual world) embeds the data model representing that evidence (crudely, just in case it faithfully depicts the part of the world that has actually been observed.)

Consider a pair of empirically equivalent theories, M and N, and pair up each model in M with its empirically equivalent counterpart in N. It should be clear that the higher-level structure of the models in each pair may differ, indeed, may differ as much as you please. This is just a mathematical fact about models; their lower-level structures may match perfectly while they differ with respect to higher-level structure. Does this entail that empirically equivalent *models* can represent the world differently, and hence that empirically equivalent *theories* may differ in content? No. Just as models which have very different structures may both accurately represent a single object, (e.g., a two-dimensional line drawing and a rod and ball model can both—despite ther internal differences—the same atom), so the models in each pair—even though they have very different higher-level structures—may constitute equivalent representations; they may represent the world as being precisely the same way. Whether they do depends on their interpretation, and whether they always do depends on how the higher-level structure of the models of a theory is determined in general.

There are two points to make:

(i) if the interpretation of the empirical substructures of a model determines the interpretation of the higher-level structure, any pair of empirically equivalent models is equivalent *tout court*, and hence any pair of empirically equivalent *theories* is equivalent tout court, and

(ii) if, moreover, the interpretation of the higher-level structure is fixed by 'charity', then any empirically adequate model accurately represents the structure of the world, and hence any empirically adequate theory is true. What it means for the interpretation of the higher-level elements to be fixed by charity is this; assume an interpretation of the empirical substructures, i.e. a mapping from the properties which define the lower-level structure of the model onto observable real-world properties. Now, find the empirically adequate model, the one which accurately portrays the observable structure of the actual world. The higher-level properties of the model correspond to whatever sets of physical objects are the real-world images of the model elements to which they are attributed. If you want to know, for example, what the property of having value x for a quantity q is, you simply locate the images of the elements in the model which are assigned value x for quantity q. This set of objects *is* the physical property having value x for q, and that's all we need to say about it.

Here is a picturesque illustration: suppose the world consisted only of an infinite patterned surface of green, red, and black, and consider a model of the world that is not an abstract mathematical one of the sort we are usually working with in physical contexts, but something quite concrete: say, a purple sheet floating a few hundred feet over its surface. Each point on the sheet has little flap of paper attached to it, and written on the paper is an ordered quintuple of real numbers which sum to 0,1, or 2 (we can require, if we like, that the assignments of numbers satisfy some mathematical constraints; we might require, for instance, that they vary continuously over the surface). Suppose, moreover, that we have a partial interpretation of the model in the sense that we have a mapping of parts and properties of the sheet onto the observable parts and properties of the world. Specifically, we know that any region of the world is represented by the region of the sheet which 'covers' it (i.e. a point p in such and such a region of the world corresponds to the point p' on the sheet which is intersected by a line perpendicular to the surface and extending from p). And we know, moreover, all regions of the sheet containing points whose attached numbers sum to 0 correspond to green regions of the world, those containing points whose numbers which sum to 1 correspond to red regions, and those containing points whose numbers sum to 2 correspond to black regions. Now, if we want to say what physical property corresponds to the sheet-property 'having a paper attached whose first number is .0017', or—better—'having value .0017 for quantity q1', we just find the physical points covered by parts of the sheet which have attached papers with .0017 written in the first place of the n-tuple. This set of points is the physical property of having value .0017 for quantity q1. Similarly for the values of other quantities.

There are several things to notice; first, the observable properties of greenness, redness, or blackness will not necessarily correspond to regions of the world which share values of any of the physical quantities. Greenness, for example, will be a complicated, infinite disjunction of physical quantities (all the different ways in which an ordered quintuple of real numbers can sum to 1). Nor will the sets of objects which share the value of a given physical quantity have the same color. In fact, there will not in general be a way to define the value of a given physical quantity except by completely describing the model, anchoring it to the world at points by giving the partial interpretation, and saying that the value of the quantity is just the set of physical points which are covered by regions of the model to which the value is assigned. That is to say, the only way in general to define the values of the physical quantities in terms of observable properties is by using the theory itself as an implicit definition of the quantity. The property of having value x for q_1 is just the property which plays the role of having value x for q_1 in the theory, the property which bears such and such (mathematically describable) relations to other values of quantities which are jointly responsible for such and such observable effects.

Second, if the sheet-model is indeed empirically adequate, and the interpretation of its higher-level properties is determined by charity, there is no way for it to *mis*represent the world. We can see quite vividly, in these terms, that any pair of empirically adequate models, both of the higher level structure of whose models is charitably interpreted, will turn out to represent the world in the same way, *viz.*, as it actually is. Suppose, for example, that we assign new quintuples of numbers to each point on our sheet in such a way that each number in the quintuple is different from what it was in the original model, but they sum to the same as they did there. The new model, different as it is from the old, if it is interpreted charitably, represents the patterned surface just as accurately. Where the two differ is in the interpretation of the quantities whose values are represented by the quintuples characterizing the points; the two differ, for example, on just what sets of points correspond to—for example—having value .0017 for q_1, the two differ, that is to say, on what "q_1" refers to. To allow that empirically equivalent theories differ in content, we must be able to vary the interpretation of the higher-level elements of the models, independently of the unobservable aspects of the world, while holding fixed the interpretation of the lower-level elements and the world's observable structure. This is just what we can't do this if the interpretation of the higher-level elements is determined charitably, for in that case what varies as we move from one world to another, is not the representational accuracy of the model but its *interpretation*. One way to think of it is that the model represents the distribution of the values of some undefined set of quantities, we find out what those quantities are by finding out what quantities are actually so distributed. So the theory provides a kind of uniquely identifying description it own basic quantities; what the quantities are depends on the way the world is; it depends on what, in the world, actually satisfies the description. So long as the interpretation of higher-level structure is determined in this way, the representational accuracy of the lower-level elements in a model guarantees its representational accuracy with respect to the higher-level elements, and the empirical adequacy of a theory guarantees its truth.

Consider the relation between the properties of the model and physical properties they represent. The model itself has all sorts of properties (it may be made of patches of different types of material, it is purple, maybe its temperature varies from one section to the next, and so on), but we pick out the properties we are interested in, i.e. those we will imbue with physical significance, by hand. In this case, it is the set of properties 'having a paper attached whose nth number is r', or, for short, 'having value r for quantity q_n,' (where r is a real number and n is an integer between 0 and 6). Notice that this set of properties has an internal structure; values of different q_n are independent of one another, different values of the same q_n are mutually exclusive and well-ordered, and so on. The interpretation hooks up certain of the properties in the one set with certain of the properties in the other, so that aside from the internal relations between properties within each set, there are external relations *between* the two. We can think of the properties in the model as having an internal structure which forms a network that gets nailed to the physical world where the interpretation hooks up one of the properties in the model with a real-world property. The interpretation of model-properties which are not themselves nailed to the world is determined charitably, and the whole effect is to bring into relief the pattern formed by the actual distribution of certain quantities. Think of shining a light on a patterned surface that turns everything black save for regions painted in fluorescent green. A slightly better analogy is provided by those 'magic eye pictures' which have a picture of some simple scene embedded in a very complex repeating pattern of colors. At first, the scene is altogether hidden, all that one sees is the complicated chaotic surface pattern, but if one focuses on the right quantities, attends only to differences in *their* values, ignoring other, the picture emerges. So long as those of its properties which are 'nailed to the world' are the observable ones and its higher-level properties are ascertainable, any empirically adequate model will provide the makings of a uniquely identifying description of the higher-level properties in terms of the lower-level ones: describe the model, describe the world, say where the world is anchored to the model, and let the higher-level properties be determined charitably, as above.[32]

The point is general; consider any two classes, **A** and **B**, of model-properties, if the properties in the **A**-class are nailed to the world and each of the properties in the **B**-class is uniquely related to one in the **A**-class, then we can let the interpretation of the properties in the **B**-class (i.e. the real-world properties they represent) be determined charitably. Whenever we do so, the representational accuracy of the model with respect to the **A**-properties will guarantee its representational accuracy with respect to the **B**-properties. The same point can be made metalinguistically, and in the special case in which the **A**-properties are observable, and the **B**-properties unobservable ('theoretical'), what we end up with is the Ramsey-sentence view of the reference of theoretical vocabulary.

---

32. *Modulo* cardinality constraints. The alert reader will also have noticed that I am ignoring the possibility that the network itself is symmetric with respect to the interchange of any set of properties.

(b) interpretation on syntactic view; Ramsey-sentences

Before I say how this works, let me say something about the notion of interpretation because, like the related notion of a model, it is often remarked that the way the term is used in association with the semantic and syntactic conceptions of theories is different. Properly conceived, I think they are the same, but it takes a bit of work to conceive them properly. As we know, on the semantic view, a theory is identified with a set of mathematical structures, its models, and the 'interpretation' of the theory is the mapping from aspects of the models onto aspects of the physical world. Specifically,

{model properties} ——*interpretation*——>{physical properties}

R   R

{model individuals}——*interpretation*——>{physical individuals}

Where R is just the relation born by the elements in the top set to those in the set underneath, *viz.*, 'being the property of'. A model correctly represents the physical world iff the two have the same structure under the intended mapping (i.e. *iff* any model individual has a given model property P just in case its image bears the physical correlate of P under the intended interpretation).[33] That's what an interpretation is and what it means to say that a model correctly represents the world under the intended interpretation. The question of how the interpretation is established is a separate one, and I suggested above that it works in the following way: we pick out the set of representing model properties 'by hand', stipulate that certain of them shall represent observable properties, find the model property which represents the actual world, and let the higher level properties of the model be deter-

---

33. I include n-tuples of individuals among the individuals (though not the basic ones) and regard n-place relations as properties of n-tuples.

mined charitably so that the interpretation of higher-level properties is determined jointly by their relation in the model to observable properties and the way the world is actually arranged.

I'll capitalize the first letter of 'interpretation' when I mean to use it to refer to the notion as it is used in model theory, to distinguish it from the more general one I have been employing. What has interpretation, in this sense, to do with Interpretation in the model-theoretic sense? A set of sentences is a structure of a certain sort. An Interpretation of such a set (or better, the language in which they are written) is a specification of a set of individuals which act as a domain, a mapping of singular terms in the language onto individuals in the domain, and of predicates onto properties (i.e. sets) of those individuals.

{predicates}————-*Interpretation*———->{physical properties}

R*   R

{singular terms}———--*Interpretation*————>{physical individuals}

R*, here, is the relation born by a predicate to a singular term to which it applies and R is—as above—the relation 'being the property of'. Instead of concentrating on the sentences describing the laws of a theory, as is usually done, concentrate on a certain class of sets of sentences picked out by these; those consistent, deductively closed, maximal sets of sentences which include the laws. Each of these provides a kind of linguistic model of a physically possible world, a maximally specific description of a world which satisfies the laws, and there is no world description satisfying the laws which is not in the set. Now the parallel to the semantic view is clear. Any of the linguistic models accurately depicts the actual world just in case: it contains a sentence of the form Pa *iff* the image of a under the intended Interpretation has the property represented by P.

In both the cases of these linguistic models and the mathematical models associated with the semantic view, we have a structure: a linguistic structure, formed by a set of words and their logical properties, in the one case, a mathematical structure, formed by a set of mathematical objects and their mathematical properties, in the other. In both cases, the interpretation is a mapping from elements of the structures onto physical elements, and in both cases, the structure correctly represents the physical world just in case the two are isomorphic under the intended I(i)interpretation. The difference between them is that in the case of mathematical models, the relation between the elements of the model which represent physical individuals are themselves properties of the elements of the model which represent physical properties. Not so in the linguistic case; a predicate is not a *property* of a singular term, with respect to the diagram above, R* is not the same as R. But this is not a relevant difference; all that matters for the purpose of using one kind of structure as a model for a physical situation, say with structure W, is that there be two sets of features of the model, and *some* relation R# between the elements in the two sets, such that the configuration of elements with respect R# is W.[34]

(c) charity and Ramsey-sentences

Consider the set of linguistic models in the above sense, and divide the non-logical vocabulary of the language L in which the sentences are written into two classes: terms which refer to observable things, and properties and terms which refer to unobservable things and properties: 'observational' and 'theoretical' vocabulary, respectively.[35] The structure of the observable world

---

34. Not just any old set of features of the model will do. So, for example, if we let pairs of real numbers stand for points on a surface in physical space, we can use the differences between them to represent distances on the surface because the set of possible differences between such pairs has the same structure as the set of possible distances between points. The same is not true of ordered pairs of rational numbers, so we couldn't use differences between *rational* numbers to represent distances in physical space. Colors can be used to represent altitudes on a topographical map because the relations between colors (nearness on the color spectrum) have the same structure as distances in two dimensions. Colors couldn't, however, be used to represent distances in three dimensions. Linguistic items like predicates can be used to represent properties, because the logical relations between predicates reflect the algebraic relations between the properties they represent.

35. As I remarked in section 3, once we have distinguished the observable from the unobservable properties, we can divide the terms in a theory's language into observational and non-observational vocabulary. The truth in the claim that the observable/unobservable distinction is not a syntactic one is only that there are no syntactic markers which distinguish observational from non-observational vocabulary, and so there is no workable syntactic criterion for separating the observational from the non-observational vocabulary.

is described by the sentences containing no theoretical vocabulary, these comprise the lower-level structure of the models. The sentences containing observational and theoretical vocabulary describe the intra-model relations between higher and lower-level structure, and the sentences containing only theoretical vocabulary describe the higher-level structure. We assume that the lower-level elements (the observational vocabulary) is already interpreted, and interpret the higher-level elements (the theoretical vocabulary) charitably, thus: find the linguistic model which depicts the actual world, replace the theoretical vocabulary with free variables, and let the intended Interpretation of the theoretical vocabulary be that extension of the Interpretation of the observational vocabulary under which the world satisfies that linguistic model. The interpretation of non-observational vocabulary is thus jointly determined by that of the observational vocabulary and the way the world is actually arranged. Theoretical terms are not explicitly defined in observational terms, but their extensions are picked out by a (rigidified) definite description statable in observational terms, which is to say that the theory provides an implicit definition of them in observational vocabulary.[36] Those who are familiar with it will recognize this as the Ramsey-sentence view of how the Interpretation of theoretical terms is determined. Gone is the reliance on the objectionable positivist version of the observational/theoretical distinction, and some of the other old positivist trappings, but the heart of the idea is the same. The view that higher-level structure is determined charitably is, in essence, the Ramsey-sentence view of the Interpretation of theoretical vocabulary transposed into the setting of the semantic view of theories.

(d) reasons for thinking higher-level structure is interpreted charitably

So, the question of whether empirically equivalent theories differ in content turns on the question of how, in semantic terms, the interpretation of the higher-level elements of a theory's models is determined, or, in syntactic terms, the Interpretation theoretical vocabulary is determined. If the interpretation of higher-level structure is determined charitably, empirically equivalent

---

36. I won't worry for the moment about details: the conditions under which the description is uniquely satisfied, what is the intended Interpretation in case there is no unique satisfier, etc.

theories are equivalent *tout court*. Do we have any reason for thinking that the interpretation of higher-level structure is determined charitably? Yes. If what I said about the observable/unobservable distinction in the previous section was right, then we don't pick out the properties which determine the underlying structure of the physical world by their intrinsic natures; the sets of physical objects (observable and unobservable alike) which figure as the values physical quantities of our theories are united *only by their external relations to particular ways of appearing to us*, where these natural types into which ways of appearing fall are taken as primitive. The observable properties are those tracked by kinds of perceptual states, and the unobservable properties are picked out by their relations to these, given by the intra-model relations between lower-level and higher-level properties.[37]

5. WHY DO WE DISCRIMINATE BETWEEN EMPIRICALLY EQUIVALENT THEORIES?

(a) Theoretician's Dilemma. *Cognoscenti* will recognize that, in denying that empirically equivalent theories can differ in content, I have grasped the more unpopular horn of what is known as the Theoretician's Dilemma. The dilemma is that if, on the one hand, one holds that all empirically equivalent theories are not equivalent *tout court*, one must say how we choose between them, since our evidence is all empirical. If, on the other hand, one holds that all empirically equivalent theories are equivalent, one must say what, precisely, the purpose of higher-level structure is.

The difficulty for those who grasp the first horn is to give reasons for thinking that the criteria employed for choosing between theories has a reliable connection with truth, without attributing us *a priori* knowledge of the relation between observable and unobservable features of the world. The epistemically most conservative choice among a set of empirically equivalent alternative theories is the theory which says as little as possible beyond what is

---

[37]. This view is quite closely related to the view that Russell apparently held in *The Analysis of Matter*, and it is useful to discuss it in connection with an objection to Russell raised by a mathematician named Newman, which Friedman and Demopolous have brought back into discussion ("The Concept of Structure in *The Analysis of Matter*", in *Minnesota Studies in the Philosophy of Science*, vol. X, p. 183-199), but the historical details are a bit tangential to our main concern.

given directly in experience, yet our physical theories are getting ever farther from the straightforward phenomenological description, we are embedding in ever higher-levels of structure. Are we making wild inferential leaps to conclusions which we can never check? And on what basis?[38]

The burden for those, like myself, who grasp the second horn of the dilemma lies in dispelling the appearance of incompatibility between theories which differ only in higher-level structure and in answering two questions:

(i) *How* do we choose between empirically equivalent theories? and
(ii) *Why* do we discriminate between them?

If all empirically equivalent theories are equivalent, then all are equivalent to the theory which contains no higher-level structure at all and consists only of a qualitative description of the world, i.e. a description in terms of the properties in which the world presents itself directly to our senses. Why do we go to the trouble of translating the description first into, and then back from, a description in terms of theoretical quantities?

All of what preceded has been prelude to answers to these questions. We needed to make the notion of empirical equivalence precise and say something about the content of theories before we could begin to address them. The question of *how* we choose between empirically equivalent theories is a descriptive one. I will give a sketch of an answer in the next section, but I'll take up the other question, the question of why we discriminate between such theories, here.

---

38. This is van Fraassen's main argument for distinguishing acceptance from belief in a theory, and withholding the latter:

> "the virtues of explanation, insofar as they go beyond description, may indeed provide reasons for acceptance of the theory, but not for belief....to be more explanatory, the theory must be more informative...both explanation by identification ... and by unification are achieved at the cost of greater information content. But to contain more information is, to put it crudely, to have more ways of being false and, hence, to be no more worthy of credence. I assume that no one can coherently call one hypothesis less likely to be true than another while professing greater credence in it..." (*The Scientific Image*, p. 44)

(b) theories and evidence

The mathematical form in which the observational evidence for our theories is expressed are the 'data models'; they comprise only a part of the empirical substructure of *one* among the theory's models, and provide only minimal constraints even on the structure of the particular model in which they are embedded. Imagine that there were a perfect division of scientific labor, that experimental scientists gather evidence, and—at the end of each day—hand theorists a data-model which incorporates all evidence observed to date. The theorists then sit down in front of the fat catalogue of All Possible Theories, data-model in hand, and begin to shop.[39] If they cross off their list theories which contain no model with an empirical substructure which embeds the data model, infinitely many remain, and these differ from one another in either or both of two respects;

(i) the way the data models are rounded out to full empirical substructures, or
(ii) the higher-level structure of the models in which the latter are embedded.

Choices between these correspond to what are usually represented as two separate types of inference:

(i) simple induction, i.e. guessing at future observations on the basis of past evidence, but also filling in bits of past and present experience which have gone unobserved, and
(ii) inference to the unobservable structure of the world on the basis of its observable structure.

The former involves a choice between theories whose models embed different sorts of empirical substructures, but each of which possesses model which embeds the data model. The latter involves a choice between theories whose models have the same empirical substructures, but differ with respect to higher-level structure. All of the theories agree on actually possessed evidence, but only in the former case is there *possible* evidence that would tell

---

[39]. It goes without saying that this isn't description of how theory choice actually proceeds, but as a picturesque way of illustrating what the choice consists in.

between them. Only in the former case, that is, do the theories in question make different testable predictions.

(c) inference to higher-level structure

Why do we ever go beyond the evidence we possess? (At least part of) the *raison d'être* of science is to *extend* our knowledge of the observable beyond what merely has been observed to what *will be*; we have a direct practical interest in prediction and control of the observable. On the face of it, however, this doesn't help explain inferences to higher-level structure, for the troublesome thing about inference to *higher-level* structure—and the thing that distinguishes it from simple induction—is that it doesn't increase predictive strength, for the empirical content of a theory is entirely contained in its lower-level structures. The reason for making discriminations among theories which agree on lower-level structures is obscure; if such theories differ in content, it looks like we are being doxastically reckless by opting for any but the theory which contains no higher-level structure at all, the one which does no more than represent the facts *as they appear*. If, on the other hand, such theories don't differ in content, wherein lies the superiority of one over any other?

(d) lower-level induction

Let's begin by looking more closely at lower-level induction. When we engage in induction, we generalize on our experience. On the basis of regularities that hold in our experience, we form expectations about regularities which hold throughout nature as a whole. So, for example, when I throw a heavy object into the air, I expect it to come down because it has always done so before; when I put water in a freezer, I expect it to freeze because it has always done so before.[40] We reason from a regularity in the association between two properties in previous experience, e.g. the fact that all observed A's have been B's, to their continued association in the future. So, for example, we observe that this emerald is green, that emerald is green, indeed all emeralds that we have

---

40. It is not, perhaps, that we *believe* that regularities observed in the past will persist so much as that we *count on* their persistence, in the same way that the dog who turns up by his bowl at feeding time every day counts in it being full, or the bird who leaps off the stony crag counts on the buoyancy of the air.

examined so far have been green (and we have examined a good number), we form the expectation that future emeralds will be green.[41] The more emeralds we have examined in the past, and the more perfect the observed correlation, the more money we would put on the greenness of the next emerald.

The general statement: "All emeralds are green" functions as a kind of bridge between instances of observed emerald green-ness and the inferred conclusion of the greenness of the next emerald. We can either represent the inference as an inductive inference from a set of particular observed instances to some future one, in which case, the general statement is not a necessary step in any of the individual inductions but a *consequence* of the whole lot of them.[42] Or, we can represent the inference as a special kind of inference from an observed past regularity to a *law* (expressed by the general statement), and then a deductive inference from the law to particular future instance of it. In the latter case, the general statement is a necessary step in the inference; one must pass through it to arrive at the conclusion. If there is any substantive difference between these two alternatives, it won't be relevant here.

In science, we have made something of an art out of induction; we have become ever more systematic in gathering and recording experiences and ever more subtle in teasing out hidden regularities.[43] The hidden regularities describe relations between the values of quantities (different quantities at the same time, or the same quantities at different times) which hold whenever and wherever those quantities obtain, and are expressed by the equations of our theories. It is these that we regard as our most powerful and secure inductive hypotheses; they have a precision and an generality that is unmatched by the rudimentary inductions of common sense.

---

41. Let me preempt the worry that concentrating on oversimple examples, i.e. exceptionless coocurrence of simple pairs of properties like emeralds and greenness, is misleading, and that a discussion based on such examples bound to be grossly inadequate to the complex inductive hypotheses framed by our physical laws. Actual laws have the form of relations between the values of sets of quantities; instead of pairs of simple properties they relate n-tuples of families of properties, and the relations between them can be more complicated than mere cooccurence, but the generalization is straightforward, so no harm is done concentrating on the simple examples. This characterization goes for dynamical regularities, regularities in the motions of systems, as well, the difference being that these describe not relations between different properties at a single time, but positions over time.
42. By that different species of induction, the mathematical sort.
43. I take this happy way of putting it from Gideon Rosen.

It is a commonplace since Hume that—although we generalize instinctively from past experience, and do so as a matter of practical necessity—we do so with no little optimism, for experience itself gives us no reason for expecting the future to go one way rather than any other. Insofar we use past experience in a systematic as a guide in forming opinions about future experience, our expectations must be structured by a principle relating past and future experience (Russell lets it be implicitly defined by our inductive practices and dubs it the 'Inductive Principle'), but experience (so much of it as we have ever yet had) can provide no support for the Inductive Principle; no principle which bridges past and future experience can be established on the basis of past experience alone. As Russell famously put it in *The Problems of Philosophy*:

> "All arguments which, on the basis of experience, argue as to the future or the unexperienced parts of the past or present, assume the Inductive Principle; hence we can never use experience to prove the Inductive Principle...
>
> The general principles of science ... are as completely dependent upon the Inductive Principle as are the beliefs of daily life. ... Thus all knowledge which, on a basis of experience, tells us something about what is not experienced, is based upon a belief which experience can neither confirm nor confute." ("On Induction", p. 108).

Philosophers have reacted in various of the ways canvassed, here, by Salmon:

> "In the face of the impossibility of any [vindication of induction], based upon a proof of the universal or frequent success of induction, what can be done?...One alternative is to *postulate* its frequent success; this is the path Russell took. Another alternative is to deny that there is any such thing as non-demonstrative or inductive inference; that is the path Popper took.[44] Another alternative is to claim that we have an intuitive sense of inductive validity to which we can appeal for justification; that is the path Carnap took.[45] Another alternative is to say that is was a silly question in the first place; that is the path that Strawson took.[46] Another alternative is to swallow circular reasoning without boggling;

---

44. Popper, *The Logic of Scientific Discovery*.
45. Carnap, "Replies and Systematic Expositions", in Schilpp (ed.), *The Philosophy of Rudolph Carnap*, La Salle, Ill. 1964, p. 977-9.
46. P.F.Strawson, *Introduction to Logical Theory*, London 1952, chapter 9.

that is the path of Black and Braithwaite.[47] Hume took up backgammon." (W. Salmon: "Russell on Scientific Inference", *Bertrand Russell's Philosophy*, ed. Nakhnikian, p. 198)

But quite aside from any aspiration of justifying our inductive practices (as Russell would have put it, aside from any aspiration of providing an *a priori* justification of the Inductive Principle), the fact is that we do generalize on experience, and do it with startling success, and there is a serious question about how we do so. We can think of this as a problem in artificial intelligence, if we like, the problem finding how to program a computer so that when fed information about past experience, it will spit out the expectations that our actual inductive practices would lead us to form on the same basis. Many philosophers in the beginning of the century—primarily among them, Goodman—came to hold that the real constructive philosophical project concerning induction lies here:

> "Now, obviously the genuine problem cannot be one of attaining unattainable knowledge or of accounting for knowledge that we do not in fact have. ... The problem of induction is not a problem of demonstration but a problem of defining the difference between valid and invalid predictions." (*Fact, Fiction, and Forecast*, p. 65)

It was the conclusion to which Russell, increasingly convinced of the futility of trying to justify induction, also finally came, and his later epistemological work is almost entirely devoted to it. He responds to Reichenbach's request for clarification of his views on induction in the Schilpp volume on his philosophy, thus:

> "If we are unwilling to profess disbeliefs that we are incapable of entertaining, the result of logical analysis is to increase the number of independent premises that we accept in our analysis of knowledge. Among such premises I should put some principle by means of which induction can be justified."

And he continues:

> "What exactly this principle should be is a difficult question, which I hope to deal with at some not distant date, if circumstances permit." (Russell, Schilpp, ed., *The Philosophy of Bertrand Russell*, Evanston, Ill., 1944, p. 68

---

47. Max Black, *Problems of Analysis*, Ithaca, NY, 1954; R.B.Braithwaite, *Scientific Explanation*, NY 1953, chapter 8.

Human Knowledge, It's Scope and Limits (1948) is his attempt to carry out the promised investigation. It is no trivial enterprise, for induction is not all curve-fitting and guessing that emeralds will continue green; as I remarked earlier, we have made something of an art of it in science and a description of our best inductive practices is nothing short of a description of the principles which govern scientific inference, the principles (if such there be, i.e., if this a properly thought of as a principled activity) which take us from the fragments provided by experience to the full-blown theories which predict solar eclipses and radio-active decay rates.

The project early on (and in sync with the spirit of the times) took the form of the attempt to develop an inductive *logic*. The idea was to proceed on the model of deductive logic and—going by our judgments of inductive validity and invalidity—to find a purely syntactic relation which holds between the premises and conclusion of good inductive arguments, i.e. to find a syntactic relation that holds between two propositions just in case one provides 'evidence for' or 'confirms' the other. The leaders in the investigation were, of course, Carnap and Hempel, and its history is well known. That it is the history of a doomed idea was demonstrated conclusively by a simple and very beautiful example of Goodman's that forever changed the way we think about induction:[48]

> "Suppose that all emeralds examined before a certain time t are green. At time t, then, our observations support the hypothesis that all emeralds are green ... Our evidence statements assert that emerald a is green, that emerald b is green, and so on; and each confirms the general hypothesis that all emeralds are green. So far, so good.
>
> Now let me introduce another predicate less familiar than 'green'. It is the predicate 'grue' and it applies to all things examined before t just in case they are green but to [things examined after t] just in case they are blue. Then at time t we have, for each evidence statement asserting that a given emerald is green, a parallel evidence statement asserting that emerald is grue. And the statements that emerald a is grue, that emerald b is grue, and so on, will each confirm the general hypothesis that all emeralds are grue. Thus ... the prediction that all emeralds subsequently examined will be green and the prediction that all will be grue are alike confirmed by evidence statements describing the same observations.

---

48. The idea was 'doomed' only taken on its own terms, as providing purely syntactic criteria for inductive validity. It can play a role in a less ambitious project.

But if an emerald subsequently examined is grue, it is blue and hence not green." (Goodman, "The New Riddle of Induction", in *Fact, Fiction, and Forecast*, p. 75)[49]

Let me introduce some terminology: I'll call the distribution of a set of properties **characteristic** just in case its distribution in the region of the universe we have experienced is characteristic of distribution throughout, i.e. just in case it constitutes a representative sample of its distribution throughout the universe as a whole. Only relations between properties whose distribution is characteristic are projectible; this follows from the meanings of 'projectible' and 'characteristic'. The direction in which we generalize on experience is determined by the set of properties whose distribution we take to be characteristic, and so even *before* we can apply any Inductive Principle we need to choose a particular set of properties and apply it to associations between those, for the Principle (insofar as it is a purely syntactic one) will deliver different verdicts depending which properties we choose. Of course, the *right* properties (where the test of rightness is predictive success) are the ones whose distribution is *in fact* characteristic, and the problem of induction is that this is something we cannot know on the basis of pre-t experience alone. One wants to say, faced with Goodman's example, that what's funny about grue/bleen properties is that if we attend to them, we will end up having to say

---

49. Two conflicting definitions of 'grue' and 'bleen' have been given (sometimes inadvertently) in the literature, and different reactions are appropriate depending on which is in play.

grue$=_{def}$ green if being examined before Jan. 1, 2000, blue if being examined afterwards.

bleen$=_{def}$ blue if being examined before Jan. 1, 2000, green if being examined afterwards.

grue*$=_{def}$ green if first examined before Jan. 1, 2000, blue otherwise.

bleen*$=_{def}$ blue before Jan. 1, 2000, green afterwards.

To know whether an object is grue, one needs to know what the current date is, and what color the object is. To know whether an object is grue*, on the other hand, one needs to know what color it is and the date on which it happens to have been examined for the very first time. So, in the case of grue*ness there is no local observational test for the property (one can determine for certain if it was examined before Jan. 1, 2000 by observing it being examined before the appointed date, but one cannot determine that it has not been examined before Jan. 1, 2000 without waiting until afterwards, and checking with all possible observers). As a matter of fact, we don't typically project relations between properties like this for good pragmatic reasons: they are very hard to discover, and we assume they will supervene on the ones for which there is a local observational test. Even if we restrict our attention to the latter, the problem brought out by grue, remains; it is the more important one, and the one Goodman meant to be pointing to.

that an overwhelming number of objects, come Jan. 1, 2000, will spontaneously (and, by the lights of our current theories, miraculously) change from grue to bleen and from bleen to grue. Blue and green, by contrast, are stable; the overwhelming majority of blue and green objects will remain so through the turn of the century. This is correct, of course, but it misses the point, which is precisely that the difference between the two emerges only *in retrospect*. it is not something that we can know before Jan. 1, 2000, at the time that we have to make our induction, at the time we are trying to decide whether to generalize on green or grue.

The so-called New Problem of Induction, the problem that Goodman's example pointed us to, is the problem of picking, on the basis of pre-t experience alone, the properties whose pre-t distribution reflects their post-t distribution. Goodman doesn't add to the old problem of induction so much as displace it, for there is a mathematical guarantee that there is always *some* set of properties—infinitely many, if there are infinitely many things in the world—to which an Inductive Principle is rightly applied, but the real problem is that of picking out *which properties these are and picking them out on the basis of the temporally proscribed evidence we have*. A description of our inductive procedures must give an account of how we identify the characteristically distributed properties, properties whose relations to one another are projectible, on the basis of nothing but pre-t experience.

(e) ostensive definability

Goodman is usually credited with the discovery, but Russell was aware long before him, of the problem posed by grue-style predicates, and he had an answer. He put the problem a little differently, in the following way: if we have examined the first n members $a_1, \ldots a_n$ of a class A and found all of them to be members of B, we may wish to predict that the next member, a_{n+1}, will also be a member of B. If, however, we allow complete latitude in the choice of the class B, B may include only $a_1, \ldots an$, and nothing else, or it may include everything in the universe except a_{n+1}, or anything in between, and we get different inductive conclusions depending on which we choose. His response to the problem was in the spirit of Goodman's, holding with them that the Inductive

Principle can be applied only to certain special relations between classes, but his criterion of specialness was different.

Russell distinguished two types of classes, those whose extensions can be picked out by description, and those whose extensions can only be picked out with a reference to particulars. The former he called 'intensionally definable' and the latter 'merely manufactured' classes. Which classes are intensionally definable depends on the available descriptive vocabulary, and so Russell's views on the definition of general terms is relevant. These, according to Russell, can be defined in either of two ways: ostensively or by verbal definitions in terms of ostensively definable predicates. Verbal definitions are what you would expect, a new predicate is introduced by identifying it with some typically logically complex phrase containing only logical vocabulary and ostensively definable descriptive terms.[50] Ostensive definitions are the more interesting case, and they're supposed to work roughly as follows; positive and negative instances are displayed and the rest of the extension is indicated as those which are more similar to the positive than the negative instances. Russell's ostensively identifiable classes are—in my terminology—qualities; their defining characteristic is qualitative similarity, and I'll say more about them in section 5.7. The distinction between intensionally definable and merely manufactured classes coincides with the distinction I made earlier between ascertainable and unascertainable properties.

Carnap was one of the targets of Goodman's attack, and his response had the same form as Russell's.[51] He called Goodman's grue-style predicates 'positional' because of the reference to particulars in their definitions, distinguished them from the purely qualitative predicates like 'green', and answered Goodman by simply tacking onto his syntactic definition of the confirmation relation, the requirement that the primitive predicates of the language in which it is given be logically simple. Purely 'qualitative' properties (e.g., green) are logically simple because they can be expressed without the use of individual constants; positional properties cannot.[52] Goodman, of course, responded that while grue is

---

50. Or descriptive terms which themselves receive a verbal definition. Any such definition, however, if we keep replacing verbally defined descriptive terms with their definitions, will eventually lead us to one whose descriptive terms are all ostensively defined.

51. Also, more recently and in a related context, David Lewis made to Putnam. The relevant paper of Carnap's is "On the Application of Inductive Logic" (1947), and of Lewis' "Putnam's Paradox", *Australasian Journal of Philosophy*.

Science and Symmetry   157

positional in a language which takes 'green' as a primitive predicate, and green is positional in a language which takes 'grue' as a primitive predicate (just as 'grue' can be defined in terms of 'blue' and 'green' only by including a reference to Jan. 1, 2000 in the definition, 'green' can be defined in terms of 'bleen' and 'grue' only by including a reference to Jan. 1, 2000 in the definition), and, since Carnap didn't say any more, is widely regarded as having got the better of the exchange. To make the response work, Carnap needs a *non-syntactic* criterion for distinguishing a primitive descriptive vocabulary, from which the rest is thought of as built up by verbal definition; in non-metalinguistic terms, he needs a criterion for picking out a privileged set of classes or properties to function as the denotations of primitive predicates, and if he had one, he kept it to himself.

Although Carnap didn't have a ready answer to the question of what the non-syntactic difference between grue and green was supposed to be, Russell did, and it was an independently motivated distinction built into the very foundations of his epistemology. He had argued repeatedly and soundly that not all of the descriptive vocabulary in a physically significant language can be defined verbally; some of it has to be taken as primitive and these get their meanings in ostensive definitions. The extension of 'green' can be introduced by exhibiting positive and negative instances of greenness because green things *look alike*. Not so, grue; we cannot define 'grue' without referring to Jan.1, 2000, because grue things observed before Jan. 1, 2000 look different than those observed thereafter.[53] The difference between ostensively definable and non-ostensively definable properties is precisely the sort of non-

---

52. Carnap made a further distinction between *purely positional properties* (e.g. 'x=Jan. 1, 2000') and *mixed properties* (e.g. 'x is green or x=Jan.1, 2000), but it doesn't play any role in the response. He tentatively conjectured that mixed properties, like purely positional ones were not projectible, but said that the question 'requires further investigation' (p. 146).

53. Positional predicates can sometimes be defined without explicitly naming the positional marker, but only if the marker can itself be ostended. The indication of the positional marker in such cases serves the same function as its explicit mention in the verbal definition; it signifies that the defining characteristic of the class is not mere qualitative similarity, but qualitative similarity in conjunction with a relation to the positional marker. Consider a predicate 'schmue' which applies to green objects on the right hand side of a visible line drawn down the center of a table and blue things on the left. If we start with the table covered with green and blue objects, we can convey the meaning of 'schmue' by ostending green things on the right and blue things on the left. The feasibility of the procedure, however, depends on being able to *point at the line* and to exhibit positive and negative instances of schmue-ness on both sides. Trying to introduce 'grue' by ostension of instances which all occur before Jan. 1, 2000, is like trying to introduce 'shmue' keeping the left-hand side of the table covered; any normal person would certainly end up thinking it applied to green things.

syntactical distinction between qualitative and positional that Carnap needed in order to answer Goodman.

A side remark: I said earlier that the unobservable quantities which figure in our scientific theories must be ascertainable, i.e. definable (implicitly or otherwise) in terms of observable properties. It becomes clear when we see that the observable properties are the ones that are ostensively definable, that the restriction is more than just a pragmatic one; if Russell is right about how general terms get their meanings, it is not merely because we are only *interested* in experience that we restrict ourselves to describing distributions of ascertainable properties, we cannot even refer to unascertainable properties in a language in which we cannot list all of their instances by name. Although, given a set of meaningful general terms, we can always introduce new predicates with verbal definitions (implicit or explicit), we can not *extend* the expressive power of a language beyond the range determined by the properties that can be defined ostensively, or verbally in terms of ostensively definable ones, and so we cannot introduce terms which will refer to unascertainable properties (Russell's merely manufactured classes). The defining characteristic of the extension of a descriptive term is qualitative similarity, simple qualitative similarity in the case of ostensively definable terms, or something more logically complicated in the case of predicates defined verbally in terms of these. If Russell is right about how general terms pick up their meanings, our languages don't contain which denote unascertainable quantities.

Let's return to Goodman's problem. Was Carnap's response right, so far as it went? Can the whole story be that only relations between purely qualitative properties (in Carnap's sense)[54] are projectible? Surely not. If there are projectible relations between qualitative properties, then there are projectible relations between positional ones. Given a set of laws between qualitative properties, it is a straightforward mathematical exercise to formulate new laws between positional ones which hold *iff* the old ones do. If relations between green-ness and emeralds and blue-ness and sapphires are projectible, for example, then so are relations between

---

54. That is, properties which can be defined without explicit reference to particulars in a language whose primitive predicates refer only to ostensively definable properties. The properties that I have been calling qualities are simply the subset of these that are themselves ostensively definable; they don't include properties which can be defined in terms of them.

grue-ness and emerises (emeralds before Jan. 1, 2000, sapphires after) and bleen-ness and sapheralds (sapphires before Jan. 1, 2000, emeralds after). Nor is it the case that we only apply an Inductive Principle to regularities between properties that are obviously qualitative; our very best inductive hypotheses, those encapsulated in the equations of our physical theories, involve relations between properties like energy and quark color which are a far cry from blue or green.

So let's look more carefully. The fact that some association between quantities q_1 and q_2 has occurred without exception a great number of times in region R of the universe, raises the probability of its occurrence outside R only if the distribution of q_1 and q_2 in R is characteristic of its distribution *without*.[55] We must treat the tiny universe of our pre-t experience of q_1 and q_2 as a representative sample of its distribution throughout, if we are to project relations between q_1 and q_2. Only in this case does the likelihood that the association is coincidental decrease as the number of experienced instances of association increases, and hence only in this case is it reasonable to expect an observed regularity to persist. Goodman's example relies on the fact that the distribution of ever so many quantities in the region of the universe of which we have experience is quite uncharacteristic, and hence the relations between ever so many quantities in the region that we have experienced will not support generalization to regions outside it, and in particular will not support inductions to regions which include future experience. We get different and conflicting predictions about the color of the next emerald depending on whether we regard pre-t relations between *grue* and emeralds as characteristic of their post-t relations, or whether we regard pre-t relations between *green* and emeralds as characteristic of their post-t relations. The *right* set of quantities to attend to is one whose distribution in the region of the universe which we have experienced *is in fact* characteristic of their distribution in the whole, and the practical problem we are faced with in deciding how to generalize on experience is that of recognizing these as such on the basis of pre-t experience alone. How do we do it?

---

55. Throughout the universe as a whole, or only in some subregion in which the event described in the conclusion falls. It doesn't matter and I'll omit this qualification from now on.

Here, in summary fashion, is what I take the correct answer to be; it will be spelled out in the next section. Carnap's response to Goodman was —although incomplete—essentially correct. If we let the 'logically simple' properties, the primitive descriptive vocabulary in the language in which his definition of the confirmation relation is given, be those which are ostensively definable, and we allow implicit definitions as well as explicit ones, Carnap's purely qualitative properties correspond to my ascertainable ones. *What we want* is a set of ascertainable quantities forming an adequate partition whose distribution is characteristic, and which exhibit a great deal of regularity. We want the quantities to be *ascertainable* and *adequate* because only these have the right kind of relations to experience (and we can focus on the *locally* ascertainable properties, those which can be ascertained by a local test on the system in question, because presumably the others will supervene on these).[56] We want their distribution to be *characteristic* so that we can base successful inductions on observed regularities between them, and we want their distribution to exhibit a high degree of *regularity* because the more regular it is, the more powerful an inductive basis it will provide.

What we do is suppose the structure of the observable properties is characteristic, embed the data models described in terms of these in models with higher-level structure which is as nice as can be (i.e., the one with as uniform and regular a *deep* structure as is compatible with the surface variety and irregularity), and project the regularities between these higher-level quantities. Why do we embed at all? because there is simply not a lot of regularity on the surface; the sheer number of observable properties is high and their distribution, irregular. They don't provide a very powerful inductive base.

That, then, is my answer to the question with which we started, the question of *why* we choose one theory over its innumerable empirically equivalent counterparts; the detour through models with nice higher-level structure facilitates lower-level induction. It was one of the really disastrous mistakes in the early history of philosophy of science to try to understand our lower-level inductive procedures independently of our practice of inferring higher-levels of structure, and to try to understand inference to higher-level

---

56. To say nothing of the foundation for the presumption. It is one that, notoriously, fails for quantum mechanical systems.

structure independently of lower-level induction. In practice, we don't make two separate inferences: *first* deciding how to round out the data models to full empirical substructures, and *then* considering what kind of higher-level structure to embed them in; in practice, the two go hand in hand, and there is no hope of understanding either in isolation from the other.

The justificatory gap which Hume first pointed to and which he thought had to be filled with a principle concerning Nature's uniformity, i.e. the gap between what we want and what we do, is still present, on this way of understanding things, but expressed a little differently. It is expressed as an assumption, unwarranted by any possessed evidence, that the structure of the observable properties is characteristic. I have allied myself with Hume in regarding it as a part of our animal nature to generalize, and I have recognized—in light of Goodman and together with Quine—that it is a part of our animal inheritance that we are inclined to generalize in a particular direction (i.e. to project green rather than grue, i.e. to presume the distribution of observable properties characteristic and to generalize on them). It is part of our intellectual nature, however, to discern hidden regularities behind the manifest qualitative ones, and science is in the business of outing them. Science, so conceived, is a product of our dual natures, based on the assumption that the distribution of observable properties is characteristic but not limiting itself to descriptions of qualitative regularities. Our animal inclinations do well by us just in case the distribution of observable properties is *in fact* characteristic; this is not something that we can know on the basis of experience, but neither is it something about which—as Hume was acutely aware—it is psychologically or practically possible to maintain sincere skepticism.

> Undoubtedly we do make predictions by projecting the patterns of the past into the future, but in selecting patterns we project from among all those that the past exhibits, we use practical criteria that so far seem to have escaped discovery and formulation
>
> (Goodman, "A Query on Confirmation", 1946)

6. *HOW* DO WE CHOOSE BETWEEN EMPIRICALLY EQUIVALENT THEORIES?

(a) introduction

If what I have said so far has been right, empirically equivalent theories, theories whose models embed the same range of empirical substructures but differ with respect to their higher-level structures, if the higher level structure of their models is interpreted charitably, are equivalent *tout court*.[57] The differences between them are differences in the quantities in terms of which they represent the world, and criteria for the choice between empirically equivalent theories amount to criteria for the choice of quantities in terms of which to represent the world. I will suggest that there are constraints on the choice: the basic quantities must be individually ascertainable and collectively adequate, and, within the bounds set by these constraints, it is made in favor of the theories with the 'nicest' models. I have already spoken a little bit about

---

57. This is not to say that any pair of empirically equivalent theories are equivalent *tout court*. Once we have an interpreted language containing both observational and non-observational vocabulary, it is easy to write down in that language, under its intended interpretation, empirically equivalent counterparts which differ substantively in how they depict the unobservable parts of the world. The point is simply that if the higher level structure of our theories' models is charitably interpreted, we are essentially using those theories as rigidified descriptions, in observational ters, of their higher level quantitie, and this means that any pair of empirically adequate theories, so interpreted, will be true, and if both true, then equivalent. The point is made nicely in an article of Horwich's ("How to Choose Between Empirically Equivalent Theories", *J.Phil.*). Horwich distinguishes theories which differ in substance from those which are mere notational variants of one another and shows that on a Ramsey/Lewis-style view about the interpretation of non-observational vocabulary, empirically equivalent theories are notational variants of one another. (We have seen why this is so when the theories in question are empirically adequate; when they are not, their theoretical terms are non-referring, and they are trivially equivalent).

adequacy and ascertainability. A choice of basic quantities is adequate just in case the partition it defines embeds the partition defined by the observable quantities; an individual quantity is ascertainable just in case it is observable or measurable. Niceness is more complicated; I hinted at the end of the last section that it has something to do with regularity, and with providing a powerful basis for induction, and I will try to spell this out somewhat in what follows.

(b) adequacy

The graininess of a photograph determines the kinds of discriminations we can make about a depicted object. If we have an aerial photograph of a building, for instance, it determines whether we can tell how many windows the building has, whether there are people on the roof, and so on. When we take a picture, we let our interest in the depicted situation determine the level of resolution required of the film. We wouldn't require very fine-grained resolution for an aerial picture of a terrain from a low height if we want only to distinguish forested from unforested regions, but if we want to count trees and distinguish their types, we need film of a higher resolution.

A quantity characterizing elements in a set **S**, will divide S into equivalence classes so that the class to which any element belongs is that of those which have the same value for the quantity. The set of these equivalence classes, if they are disjoint and their sum is **S**, is called a **partition** of **S**. The basic partition of a theory is the partition each of whose cells corresponds to a complete assignment of values to the theory's basic quantities. Each cell in the partition corresponds to a description at the highest level of physical detail recognized by the theory. The partition is the theoretical analog of the graininess of a photograph; it determines the number and type of discriminations that can be made in the theoretical depiction of systems that fall in its domain.[58]

---

58. In general, the partition defined by the theoretical quantities of a particular science is required to be adequate to the dependent variables of that science. This goes for the special sciences as well as physics; the basic quantities of economic theories, for example, must be adequate to prices and amounts of production. What is special about physics is that - to the extent that it aims for descriptive completeness - its dependent variables include all observable physical quantities (with the qualifications mentioned in section 6).

Adequacy is a lower-bound on the level of fine-graining of the theoretical description. A set of structures S_2 is adequate to another set S_1 just in case there is a mapping of the atoms of S_2 onto elements (atoms or constructions out of atoms) of S_1, and the partition of the atoms of S_2 is a more fine-grained version of the partition of the corresponding elements of S_1.[59] So no two systems which are distinguished by properties in S_1 are equivalent with respect to all properties of the constructions with which they are identified; all differences in terms of relevant properties (properties defining the partition) of atoms of S_2 are represented by differences in relevant properties of the right constructions of smaller elements of S_1 with which they are identified. In physics, we typically identify large physical systems with more or less complex constructions out of more basic elements, and their properties with constructions of properties pertaining to these.[60] Gases, for example, are identified with configurations of molecules and properties of gases, like temperature, are identified with properties of such molecular configurations, like mean kinetic energy. The partition of the more basic elements, together with the rules for combining them into complex systems, provide a partition of the latter.

Just as in the case of a photograph in which the resolution of the film is chosen so that differences between situations we'd like to be able to distinguish show up as differences in the type and configuration of grains in the picture, the choice and partition of physical atoms is made so that *at least* the full variety of qualitatively distinguishable physical structures can be reproduced by different configurations of their smaller constituents. In the same way that if we want to be able to predict the number and type of trees in a forest from a photograph of it, the grains of the photograph should be small and various enough that the pattern of grains in a picture of 110-tree softwood grove is different from that of a picture of a 112-tree hardwood patch, the atoms of our physical theories should be at least small and various enough that the theoretical descriptions of qualitatively distinguishable systems differ. In technical terms, adequacy is the requirement that the qualitative structure of any world is a substructure of its theoretical structure; it ensures that complete information about the structure of

---

59. 'Atoms' is taken in its traditional sense, to refer to the basic unstructured constituents of matter (if any there be).
60. We typically do, but needn't. See section 6 (g).

appearance within any spatiotemporal region R is recoverable from a full theoretical description of R. Qualitatively distinguishable systems should fall into different cells of the basic partition.

We will see later that theoretical benefit accrues to a coarse-grained partition of basic elements, and coarse-graining while remaining adequate is achieved by making the atoms smaller and deriving variety at the global level from variety in construction. In the same way that we must have long words to obtain a large vocabulary from a short alphabet, and in the same way that we need fewer English letters than roman numerals to get any number of distinct postal codes, if we have just a very few *types* of basic elements we need many of them to achieve the requisite variety. Variety can only be achieved by combining them in different ways, and as the number of different combinations increases only with the number of elements, this means regarding qualitatively characterized objects as composed of a very great number of very small theoretical elements.

An analogy; think about how one learns to *hear* music, western orchestral music, for instance. At first, every piece sounds different; gradually, one is able to distinguish the sounds of various instruments and to discern that different (synchronic) sounds come from different combinations and strengths of the same set of instruments; violins, trumpets, flutes, and so on. A little more listening, and one teases out the strands coming from different instruments and finds certain phrases and themes repeated, sometimes in a similar and sometimes in a transformed manner (slowed down or reflected, for example), within the string of sounds coming from a single instrument or across strings associated with different instruments. Finally, each composition is seen as constituted out of different configurations of a small stock of types of elements. A representation in terms of a particular set of note types is adequate just in case the full variety of compositions (individuated by score) is representable as different configurations of notes.

The analog of qualitative structure is the heard structure of a single composition, and the analog of theoretical structure is the representation in terms of notes. (We could imagine, to extend the analogy, that differences in tone, etc., of individual notes is indiscernible to ordinary listeners without some special auditory implementation). The adequacy of theoretical to qualitative structure is the requirement that no qualitative distinctions are lost in going to a theoretical description, just as the adequacy of the note-representation to the variety of compositions entails that no distinct compositions get represented by the same configuration of notes. It entails that no systems which fall in the same cell of the theoretical partition are qualitatively distinguishable (otherwise the qualities which distinguish them would have the same theoretical description), and this in turn means that the qualitative structure at any world must be embeddable in the theoretical structure.

(c) ascertainability

Whereas adequacy places a *lower bound* on the level of fine-grainedness of the theoretical partition, ascertainability constrains which quantities can be used to define it and thereby places an *upper bound* on it. It is the requirement that the basic quantities be observable or measurable, i.e. that there be an operational test for deciding which cell of the theoretical partition an arbitrary physical system falls under, a test with one or another *qualitatively different* outcome depending on which value of the quantity in question obtains. The existence of operational definitions for all of the theoretical properties means there is no absolutely undetectable structure, so that if we know the definitions and are given unlimited time and experimental resources, we can determine the theoretical structure of any world from its qualitative structure. It ensures also that we can determine from the theoretical structure of a world (or any subregion of any world), information about its qualitative structure. Without a link between qualitative and theoretical properties of the sort required by ascertainability, detailed knowledge of the theoretical structure will be useless for conveying information about its qualitative structure; only if the theoretical partition is ascertainable can we retrieve information about the structure of appearance from a description of the theoretical structure of the world, and only if it is also adequate can we retrieve *full* information. We don't represent the world in terms of some partition with respect to which it has a perfectly homogenous structure, i.e. in terms of some partition such that all actual events fall in the same cell, not because there is no such partition, but because its cells correspond to unascertainable properties, and a description of the world in terms of it is uninteresting, it tells us nothing about the qualitative structure of world; it tells us nothing about how we will experience the world.

Whereas adequacy is a pragmatic constraint, ascertainability is at least an epistemic one, perhaps a pragmatic one, and possibly a semantic one. **Epistemically**, all we can determine through experience is the qualitative structure of the world, and so the only theoretical distinctions we can make are those that can be recognized by their qualitative manifestations. **Pragmatically**, insofar as the purpose of science is to extend our qualitative knowledge beyond the purview of what is given directly in experience, theoretical distinctions without qualitative manifestations are useless. **Semantically**, the only meaningful similarities and differences are those that can be qualitatively defined. The conceptual anchor of informativeness for us in a physical description, is its implications for qualitative structure. We know what to expect if we are told that no green things are heavy; we don't know what to expect if we are told (to use an example of Feynman's)[61] that all foos are schloos (at least until and only insofar as we can give a qualitative description of foos and schloos), until we are told how to recognize such things, should we come across them.

We can illustrate these points by looking at examples in which the structure of a section of the ceiling of the Sistine chapel is represented by a set of ascertainable and unascertainable properties, respectively. Consider a table-top model of the pattern of colors over a section of the Sistine constructed for a blind girl named Greta. Points on the model are mapped onto points on the relevant section of the Sistine, the 2-dimensional spatial relations between points on the Sistine are mapped onto appropriately scaled versions of themselves applying to the model, and we use the color spectrum to define a mapping of colors into two-dimensional space. The model is constructed by raising regions on the table-top corresponding to regions of a given color on the Sistine to the height associated with that color so that the pattern of heights over the surface of the table reflects the structure of the pattern of colors in the relevant section of the Sistine. Now, if we ask about the conditions under which Greta can use the model to obtain information about the pattern of colors over the Sistine, it should be clear that so long as she can determine the structure of

---

61. *The Character of Physical Law*, p. 70.

the model with respect to the height partition, i.e. so long as the heights are ascertainable, she is in clover. And she can, so she is.

Consider, by contrast, the face of the clock on the wall to my right. It also provides an adequate model of the same section of the Sistine; there is a set of properties of the clock (sets of points on its face) which are isomorphic to its color pattern. I could tell you what the relevant properties are and tell you which of them correspond to which colors on the Sistine section if I know the pattern of colors over the latter and the mapping of points onto the clock face, but this is the *only* way I have of specifying them. Obviously, I cannot use the clock face to obtain information about the pattern of colors over the Sistine because I cannot determine its structure with respect to the modeling properties without *already* knowing the structure of the modeled ones—I cannot, that is, ascertain them by experience—so for the purposes of *obtaining* information about the latter the thing is all but useless. This is the difference between Greta's table-top model and the clock-face. She can ascertain heights over the table-top by running her fingers over it; together with the information that the pattern of heights is isomorphic to the color pattern in the Sistine section, this provides her with full information about the latter. If the purpose of a model is to convey information about structure of an object with respect to one set of properties by encoding it in the structure of another set, then we had better be able to determine what the latter is without knowing anything about the former.

The example of Greta brings out another important point: ascertainability is a *relation* to beings endowed with certain epistemic capabilities, not an intrinsic feature of a set of properties. So, for example, an actual-size photograph of the Sistine provides an ascertainable model for me but not for Greta because I can, and she cannot, determine the pattern of properties (the colors in the photograph) which reflects that of the colors on the Sistine ceiling. A model wears its structure on its face if the modeling properties are qualities (as in Greta's model), but it can also be more difficult to ascertain (as when we use a radio-wave to carry the picture on a television screen, or the grooves on a compact disk to encode the sounds it carries). In such cases, its structure with respect to the

---

62. Qualitatively simple worlds with short histories do not provide counterexamples because we are not, in such worlds, given unlimited time.

relevant properties isn't immediately apparent, but so long as they are measurable, it can be ascertained.[62]

Ascertainability and adequacy both constrain the relationship between theoretical properties and qualities. Adequacy requires that the qualitative structure in any region of any model be a substructure of the theoretical structure; ascertainability requires only that there be a qualitative test for deciding whether each of theoretical properties obtains. The relationship is asymmetric: the theoretical structure within any region of a model determines its qualitative structure, but the qualitative structure in a region (or even at a world) does not determine its theoretical structure. It is only true that, given unlimited time and experimental resources, the theoretical structure at a world can be determined from its qualitative structure.

(d) the need for criteria beyond adequacy and ascertainability

So the basic partition must be adequate, which is to say that no qualitative distinctions can get lost, in going to a theoretical description of the world, and it must be ascertainable, which is to say that there must be some set of measurements one can perform to determine which cell an arbitrary system belongs to. Do the constraints of adequacy and ascertainability determine the choice of basic quantities? Not even close. Given any adequate ascertainable partition, it is a simple mathematical exercise to find innumerable gerry manderings of it which are also adequate and ascertainable. To gerry mander a partition is to define another out of logical constructions of the properties which define it. Since the class of ascertainable properties is closed under logical composition, any gerry-mandering of an ascertainable partition is itself ascertainable, and so long as P is adequate, indefinitely many of these will be as well.[63] Adequacy applies to the properties which define the theoretical partition collectively: they must provide a partition that is fine-grained enough to represent all qualitative distinctions. Ascertainability applies to the properties individually: we need to be able to provide an operational test with qualitative-

---

63. All ascertainable gerry-manderings are ascertainable, but only some gerry-manderings of adequate partitions are adequate. The partition with just a single cell obtained by disjoining the properties which define the original, so long as it has more than one cell, for example, is not.

ly different outcomes which determines whether any given property obtains. Adequacy and ascertainability are pragmatic constraints *dictated* by the aim of using theoretical models as a basis for description and prediction of qualitative structure.

> ...And this is the way to obtain as great a variety as possible, but with the greatest possible order; that is, it is the way to obtain as much perfection as possible
>
> (Leibniz, Monadology, #58)

(e) niceness

The claim, recall, was going to be that scientific theorizing is a matter of choosing a partition with respect to which the world has a precisely and objectively characterizable sort of structure. Without constraints on the class of partitions from among which the choice is made, any such account will represent theorizing as way too easy because there are partitions with respect to which the world has almost any structure you care to describe. So ascertainability and adequacy, which constrain the theoretical partition by constraining the relation it can bear to the qualitative partition, are crucial to the account but they don't come close to determining the choice: we still need to specify the structural feature of models that is maximized within these constraints, and this I will call 'niceness'.

In section (ii), I used the example of musical structure to illustrate adequacy. Compositions were represented as configurations of notes in time; each composition was taken as the analog of a single state and time as the dimension in which its parts were arranged. The analog of qualitative structure was the heard structure of a single composition, and the analog of theoretical structure was the representation in terms of notes. Since adequacy just relates the qualitative description of a system to its theoretical description at a particular time this simple analogy was sufficient, but to illustrate niceness we need something that incorporates an analog of evolution, and for this the following will serve. Imagine that we have a set of glass beads in a range of colors and sizes, and a surface divided into squares of equal size with a dip in the center of each that holds a single bead. There are, in addition, a set of dice which are continuously rolled and which have whatever number of sides and biases you like, and the configuration of beads on the board changes with each roll of the dice.

Here, then, is the definition of niceness; the games with the **nicest** structures are those which *can assume the fewest different types of configuration and whose evolution is governed by the strongest discoverable rules*. Those rules are strongest which leave the least up to the rolls of the die, and those are discoverable which are both relatively local (i.e. relate beads whose spatial separation on the board and in time is small relative to the section of the board in view and the duration of our watch) and which describe not particular events, but invariant relations between event-types, i.e. relations which hold at all times and all places on the board.[64] The nicer the structure of a game, clearly, the more we can predict about the configuration over the whole board and its future evolution by watching a small part of it for a relatively short duration. Likewise, only physical events which occur in the tiny corner of the universe of which we have experience, and somewhat local relations between types of them, are discoverable by beings whose experience is as spatially and temporally proscribed as our own. Partitions which define patterns dominated by these local, invariant relations, that is to say, are better suited to our epistemic aspirations; they allow us to map the most of the whole of it on the basis of the part of which we have experience.

How do we go about guessing at the rules of the game? The answer is simple. We watch how the configuration on the board evolves and tease out the regularities—both manifest and hidden—between neighboring squares and successive states. What I mean by teasing out regularities will be easy to say with some regimentation; a familiar way of modeling the evolution of a physical system, S, is to represent the possible states of S by points in a *phase space* associated with S. S's state at any time is represented by some point in the space and its evolution throughout any period to a sequence of points or a trajectory through it.[65] The dynamical

---

64. Both of these properties of *discoverable* rules follow from the fact that the distinction between the accidental cooccurrence of two events, and their relation by a law emerges only with a great enough number of instances, so for a law to be recognizable *as* a law we must be in a position to view a large enough set of its instances.
65. I will sometimes speak of S's position in, or trajectory through, phase space, when I should more properly say the position of the point representing S's state or the set of points representing its trajectory.

equations that govern S's evolution pick out the trajectories in its phase space that are physically possible for S-type systems. So, for example, the phase space for a Newtonian system comprised of particles of specified type and number contains a point for each possible specification the positions and momenta of all the particles comprising S, its (q_i, p_i), and the Newtonian dynamical equations pick out the trajectories through it which describe physically possible histories for the system in terms of its changing (q_i, p_i) values. Now, if (q_1, q_2...) are quantities pertaining to S, and f is an arbitrary mathematical function, then $f(q_1, q_2...)$ will be a well-defined quantity for S. Moreover, if (q_1, q_2...) are individually ascertainable, $f(q_1, q_2...)$ will be ascertainable as well, for it can be ascertained simply by determining the values of (q_1, q_2...) and then applying f. This means that, so long as S's state is ascertainable, arbitrary functions on S's phase space will represent well-defined ascertainable quantities pertaining to S.[66]

Going back to the game, suppose beads come in a continuous range of sizes and colors represented, respectively, by real-valued quantities $s(x)$ and $c(x)$, so that the state of the board at any time is given by an assignment of ordered pairs of real numbers $<s(x), c(x)>$ to square, and the phase space for an n-square board is 2n-dimensional and contains a point for each such assignment. Arbitrary functions of color and size are well-defined ascertainable quantities pertaining to individual beads, and so are arbitrary func-

---

66. This is quite standard. The only constraint is that sometimes it is required for mathematical tractability that f be differentiable. Here are two fairly typical quotes from classical dynamics textbooks.

> "it must be agreed what the [physical quantities] are... We should therefore allow arbitrary functions of coordinates and momenta as [quantities], subject only to boundedness and, for mathematical convenience, differentiability." (*Classical Dynamical Systems*, trans. E.M. Harrell, New York: Springer-Verlag,1978, p. 5).

and,

> "The [quantities pertaining to] the system are all real functions on phase space (or they may be a subset chosen according to some criterion). "
> (*Classical Dynamics: A Modern Perspective*, New York, John Wiley and sons, 1974, p. 524).

67. Or perhaps we just watch for as long as we like and assume that the sample we observe is random, so the longer we watch, the closer we come to having an infinite random sample of possible trajectories, and the more confident we can be that we've observed all possible ones.

tions of the *configurations* of colors and sizes of beads on the board as a whole.

To formulate a theory of the game, we proceed on the supposition that it is rule-governed (we assume, that is, that the configuration of beads on the board at any time, together with the results of the rolls of the die, at least partly determines the configuration at subsequent times), and we go about uncovering the recipe by which it does so. Suppose that besides watching the die roll and the board change, we can intervene and reconfigure the beads, so

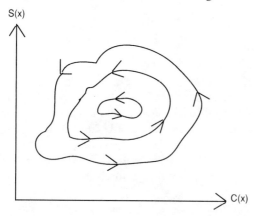

that the board will evolve naturally from its new configuration. We watch, and we reconfigure, and plot observed trajectories until we are convinced that we have observed all those that are possible.[67] Then we sit down to see what kinds of regularities the motion exhibits, and I want to emphasize is that this is not just a matter of noting invariant relationships between observable quantities. It is a matter of *smoking out hidden regularities* by canvassing the way the evolution of the board looks with respect to different sets of ascertainable quantities. Here's what I mean; focus the two-dimensional phase-space for a single square on the board, and suppose the observed trajectories of the square through its space look like the figure above.

If we made a continuous printout of such a system's changing color and size values, the results would appear haphazard and disorderly. But now notice that if we can find two quantities **U** and

---

68. The quantity u is sometimes called a constant of motion, because its value doesn't change as the system evolves.

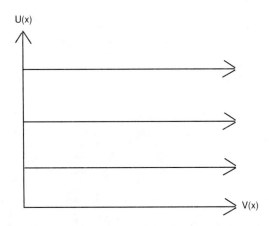

V such that S travels on a level surface for **U**, so that S's **U**-value picks out the path in the above diagram that S is on, and S's **V**-value gives its current place around that path, measured as an angle, then S's motion from the perspective of U and V is quite regular. A hidden regularity in S's motion emerges conspicuously when it is represented in terms of these quantities. Trajectories through the phase space above, written now in **U** and **V** coordinates, look like the figure shown above.[68]

The same thing will go for phase spaces of higher dimensions: consider, for example, a four-dimensional phase space and ask whether any U exists that slices the space into neat leaves so that each trajectory of the system remains on a fixed leaf of U (these leaves, incidentally, are called 'foliations'). V, as before, provides the system's current location on the relevant leaf. If there is a further quantity W which slices the phase space transversally so that each trajectory is confined to the intersection of a U leaf and a W leaf, the space is said to be 'fully foliated'.

The point of all of this is to illustrate how regularities in apparently irregular motion can emerge when represented in terms of a particular set of quantities. A choice of quantities is a good one just in case there is a particularly *simple* and transparent functional relationship between the values of the chosen quantities at different times on all physically possible trajectories. Once discerned, it doesn't matter which quantities we state the invariant relations in terms of, but the insight involved in discerning them in the first place is very much a matter of *seeing the motion in terms of the right quantities*. The most important part of theorizing is finding the quantities which bear these simple, invariant relations to one another. The idea is not unfamiliar to physicists; Bohm, for example, writes in *Quantum Theory*:

> We may ask why energy plays a more important role in mechanics than is played by other functions, such as mv^3 or arc sinh (mv). The reason is that the total energy of any isolated system is conserved, whereas, in general, no such conservation laws can be found for most other functions . . . (Bohm, *Quantum Theory*, p. 153)

Another nice example which comes from a paper by Mark Wilson (in which he also discusses foliating quantities like those above) is that of properties of metal plates discovered by the 18th century craftsman E.F.F. Chlandi when he sprinkled sand on plates held in specially clamped positions. Such plates vibrate in a combination of special patterns or 'modes' (in precisely the same way that taut strings—when plucked—vibrate in superimposed patterns of pure note frequencies). As it turns out, these mode-like properties of the plates are correlated with other important properties; they determine how the plate will absorb energy, when it will rupture, and so on. This means that a great deal of important information about a system can be had by focusing on these properties and attending to their connections with others, and they are well-placed to play an important role in theorizing. Wilson goes on to point out that:

> It turns out that a wide variety of physical systems own their own private stock of mode-like properties; many of the key successes in nineteenth-century physics represented the elucidation and exploitation of these hidden traits. The investigations of Joseph Fourier and his school (particularly those by C. Sturm and J. Liouville in the 1830s) revealed the mathematical basis for these behaviors. If a physical system is suitably modeled by a set of equations of "Sturm-Liouville type", then it "mode" properties can be calculated. If this is done, an experiment can be run—sprinkling sand, shaking the material with a frequency generator—that looks for the predicted traits.... Once recognized, the hidden

---

69. This can be relaxed; it is intended to show that such regularities as there may be between quantities at very great spatial and temporal removes from one another (at opposite ends of the universe or history) will not be recognizable as such by the likes of us because we will not have a large enough sample of quantities so separated to be able to distinguish regularities from accidents. It is probably better to call this weak requirement 'contiguity', or some such thing, and use 'locality' only in its usual sense to denote the requirement that laws relate quantities only in one another's immediate spatiotemporal neighborhood.

qualities prove most helpful in understanding the system's composite behavior.. ("Honorable Intensions", p. 67)

(f) illustrations

In the abstract terms in which I have presented it, this process of teasing out hidden regularities sounds exotic, so let me run through a couple of homey examples of it to illustrate its ubiquity *outside* scientific contexts. First, though, let me say more precisely what I mean by hidden regularities; these are local, invariant relations between unobservable but ascertainable quantities. To say they are local is to say they relate quantities at neighboring places or successive states,[69] and 'invariant' is short-hand for 'invariant with respect to spatial and temporal displacements', which means we can infer from the fact that they obtain in our experience that they obtain everywhere and always, which is to say (in the terms of section 6.5) that the quantities they relate have a characteristic distribution: they are projectible.

(i) congresswoman A. Suppose we are trying to understand the voting behavior of a congresswoman, A, who answers either 'aye' or 'nay' to each of a steady stream of questions. Judging just from her answers and the content of the questions, her behavior appears erratic; she votes differently on questions that have the same content, sometimes aye-ing, sometimes nay-ing those that are the same or are distinguished only by incidental features like wording. We might wonder whether her votes are determined not by the content of the questions, but by something else: perhaps she always follows the majority vote or the interests of a special group, or perhaps she votes 'nay' always and only if the vote takes place in the afternoon. The way to check in each case is to give her answers on questions partitioned into classes according to the votes of the relevant groups.

Suppose A's answers are random with respect to all of these, i.e., that her votes are evenly distributed across the cells of the partitions defined by each, but by examining a list of questions partitioned according to whether they were assorted or opposed by a particular congressman, B, we hit upon the fact that A ayes all and only those B nay's.[70] She is, it appears, meticulously consistent in

---

70. I'm assuming that they both vote on all questions.

opposing B's votes; far from being erratic, she votes *in the same way* every time, and we can pull off the trick of predicting her next vote as soon as we figure out what counts as the *same*. To say that A's votes always oppose B's is to say that there is an invariant relation between A's votes and B's; if we let $A(q)=$ the value A assigns question q, aye=1, and nay=-1, we can write it as $A(q)=-B(q)$.

So, B's votes determine A's, but the question of whether anything determines B's, i.e., of whether B's behavior is itself erratic, remains. What is it for B's votes to be erratic? Not simply for there to be no partition of questions such that he aye's all and only those in one cell, and nay's all and only those in another; there is *always* some such partition. It is rather for there to be no such partition of a particular kind. The relevant partitions are those that can serve as a basis for the prediction of B's vote on a given question, and what we need in order to predict B's vote is to associate each of its values with different values of something we can determine independently. A little more precisely, we need to be able to associate the values of B(q) with the cells in some adequate, ascertainable partition.

(ii) The wandering ant. Another illustration (which will be recognized by those familiar with Herbert Simon's work in AI) suppose we are trying to predict the path that an ant will follow as it searches a patch of ground for food. It wanders now this way and now that, changing its course unexpectedly and in apparently unpredictable ways. We can approximate the curve described by its path with a finite series of straight line segments of equal length, and the ant's behavior by a series of choices—one for each segment—of whether to go on in the same direction or to turn in one or another direction. Suppose, as it happens, the path describes a random curve in the mathematical sense, i.e., a curve of infinitely many degrees. Ask, now, whether the ant's behavior is completely erratic, or whether, like congresswoman A, it really does *the same thing* at every juncture? Could we, if we hit upon the relevant respects of sameness, predict what it would do in any case? Again, this is the question of whether there is some way of sorting the circumstances the obtain at junctures into an exhaustive list of alternative types $c_1, c_2 ...$, so that the answer to the question 'what does the ant do in circumstance c_n?' is the same every time, and—furthermore—which is such that which of the circumstances actually obtain at any juncture is characterizable independently of, and ascertainable without knowledge of, what the ant actually does in each case. As in the example of congresswoman A, the trick of understanding and predicting the ant's behavior is being able to associate the different directions the ant can take with different cells in an adequate, ascertainable partition of circumstances, so that we can determine independently which cell of the partition the circumstances at any juncture fall under, and use this to determine which way it will go.

As it happens, ants do follow apparently random curves, but as we understand it, their behavior is far from erratic; it seems that they follow a simple set of innate rules for responding to features of the terrain, and it is the randomness of the distribution of those features which make its path look irregular. Let's look at this example in a little more detail. The pattern in this case is a two-dimensional curve. We can separate it into a rule governed component (the relationship between the circumstances at a juncture and the direction the ant takes from that juncture), and a random component (the circumstances that obtain at a given juncture). In the best case, there is an ascertainable circumstance partition of which each cell corresponds to a single cell in the direction partition, so that what the ant does at each juncture is determined by which cell of the circumstance partition the juncture falls under. Alternatively, each cell of the circumstance partition will correspond to a probability measure over cells in the direction partition. The direction the ant takes is random with respect to a given partition *iff* each of the cells in the latter corresponds to an equally weighted distribution of cells in the former, so that the ant is equally likely to go any which way no matter which of the relevant circumstances obtain (i.e., *iff* there is no correlation between types of circumstance, and the ant's choice of direction). Less than perfect prediction of the direction the ant will take at any juncture is the same as perfect prediction of the probability with which it will take any one direction (i.e., a partition which is adequate to the assignment of probabilities where the frequencies pertaining to the occupants of any individual cell reflect the relevant probabilities).

(g) conclusion

The idea in this last section has been to try to give some content to the truism that science aims to construct theories which achieve the best combination of simplicity and strength by fixing on structural features of the models of a theory which are plausible candidates for the requisite notions of simplicity and strength, and which—moreover—make true the claim that what has been in fact regarded as scientific progress consists in an increase of the combination of the two, while making sense of science as an epistemic project. We have a great deal of freedom in the choice of a partition with respect to which to represent the physical world, and we exercise it in science to come up with a representation in terms of which the world has the nicest structure we can. But our freedom is not absolute, and the constraints as well as the purpose of the enterprise are epistemic, geared toward extending our knowledge of the world beyond the spatiotemporal reach of experience. Scientific theorizing as represented here is a way of building up our picture of the qualitative structure of the world, different from but dependent on the brute method of going out there and collecting evidence. The problem with the brute method is not that it is not a good one, but that it is spatially and temporally limited, and limited by our sensory perspective, the set of quantities to which ours senses are naturally attuned, and we are eager at any given time to know what lies yet outside our spatiotemporal reach, in the same way the blind Oedipus was grateful for the eyes of Antigone, that she could tell him what to expect to find under his feet before he chose where to lay them down.

(h) addendum

I have been speaking mostly as though science were a matter of representation and that we are concerned to fix on quantities whose observed relations can be expected to obtain in unobserved regions of the actual universe. What is missing in the picture is modality. If our knowledge of the distribution of the quantities in question is to be used for control as well as prediction, the relations between them have to obtain not only throughout the *actual* world, but throughout the larger universe of physically possible ones. This is just a picturesque way of saying that they have to get the truth values of counterfactuals right; they should tell us, and tell us correctly (whatever the relevant standards of correctness are here), not only what happens in actual situations but what *would* happen in situations that may never in fact occur (indeed, some of which we may go to a great deal of trouble, on the strength of the counterfactual, to keep from occurring).

Whatever one thinks about the metaphysics of counterfactuals, it is certain that our assessments of their truth values plays an important role in deciding how to act, and that they are very closely tied to what we take the laws to be. This is especially evident in fields like economics where the theorizing has a primarily practical import. The domain of economic theory is not so much observed as constructed; *we* construct it with an eye to achieving particular results, and our opinions about what *would* happen under a wide range conditions play an essential role in how we decide what to do. The point has been made persistently and persuasively by Nancy Cartwright;

> [econometrician's] fundamental concern is: will the parameters that have been estimated under whatever conditions obtained in the past, continue to hold under various innovations? If not, their models will be of no help in forecasting the results that *would* occur *should* the economy be manipulated in various proposed ways. (*Nature's Capacities and their Measurement*, p. 154, my emphasis)

The point is, I think, undeniable; our interest in what *would* happen is not derivative of our interest in what *will* happen; what *will* happen—which innovations we decide in the end to implement—itself depends on our judgments about what would happen. The same is true of the physical sciences, though less evidently so because a larger component of these is concerned with representation, and so much less of it is under our direct control. What it suggests, however, is that we should leave off speaking of relations between quantities which are projectible into unobserved regions of the actual world and begin to talk about relations between quantities which hold counterfactually as well. To put it a

---

71. Or so I am inclined to think. Questions about the precise relationship between causes and counterfactuals, and about the direction of metaphysical determination (to the extent that such questions make sense, are difficult, and I can avoid them. What does seem right is that the ontology of physical theories is an ontology of capacities, that the laws express causal relations between quantities, and that they have counterfactual import.

little differently, and in the terms that Cartwright herself favors, perhaps we should leave off talking about mere regularities, i.e. actual patterns of covariation between quantities, and start talking about their *causal* relations, for it is the causal relations which decide how things will go in counterfactual circumstances.[71] Or perhaps there is some alternative which would work just as well.

Cartwright's own suggestion is that in choosing basic quantities, we are trying to isolate what she calls 'capacities', parameters which are such that—vary them how you will, and vary everything else as you please—the relations they bear to one another remain both actually *and* counterfactually. If **A** has the capacity to raise the probability of **B** (or, more generally, to affect **B** in such and such a characteristic way), then it does so against all fixed backgrounds.[72] This means that if we hold everything else fixed, and we have the right kind of control over **A**, we can use it as a kind of knob to 'turn up' the probability of **B**.

> the [equations relating the quantities of a theory] express a commitment about what remains constant under change...The methods presuppose that causes have stable tendencies of fixed strengths that they carry about with them from situation to situation. What **A** contributes to **B**—its total influence, a**A**—depends on **A** alone, and is the same no matter what goes on elsewhere. (*ibid.*, p. 153, I have changed her variables to match those in the preceding paragraph).

There's an epistemological question about how, on the basis of knowledge of observed patterns of co-occurrence of actual events, we can fix on the capacities. It is the question of how we can project invariant relations in actual situations into *possible* ones as well, and how we isolate the capacities of individual properties given that they always operate *in concert* and always against one or another background. The former is the analog of the problem of induction in application to counterfactual situations, and I think it should be answered as the ordinary problem of induction, *viz.*, by laying out the principles which appear to govern our inductive inferences. It should parallel the general story that I have told about

---

72. To say that **A** raises the probability of **B** against all fixed backgrounds, is to say that the effect of **A** on **B** has *contextual unanimity*, (Cartwright borrows the term from John Dupre). Fixed backgrounds are just backgrounds against which all other causally relevant quantities have fixed values, and so the epistemology of capacities turns out to be holistic, Individual claim about capacities can't be tested without presupposing others, for we can't say what count as fixed backgrounds until we know what the full set of causally relevant factors are, and we can't say this until we know something about which factors individually influence the parameter in question.

ordinary induction. There is a more detailed question as well: given a choice of quantities and the observed patterns of coocurrence they actually exhibit, how do we infer the more detailed causal structure of those quantities, the network of chains of direct and indirect causation between them? This is the question to which Cartwright's work and that of Glymour, Spirtes, et.al. is geared, and they offer a battery of techniques (mostly statistical: some ingenious, and some simple common sense) for establishing and testing such hypotheses.

The examples I discussed in sections **6.7 (f)** and **(g)** were cases in which the state at one time was regarded as the cause of the succeeding state, a Newtonian separation of space and time was assumed, and individual states were left unanalyzed. Transposing the discussion to a proper relativistic setting will make it more difficult to find easily visualizable examples but shouldn't pose any additional problems. If we ignore the question of how the individual states are further analyzed, however, we leave out a large part of the story. Individual states are decomposed into the values of a set of quantities whose effects on one another can be cast in more or less simple functional terms, and which act together according to relatively simple rules to produce the full variety of observed effects. The picture to have in mind is that the world is like a big machine, and the job of theorists is to figure out what the individual parts can do and reconfigure them to build new ones that perform quite different tasks. Isolating individual capacities is a matter of dividing up the joint powers of various combinations of them, it is like figuring out what parts of the machine *can* act by seeing how they *do* act in concert with others.

So there, in broad strokes, is a picture of how we go about formulating theories. Our evidence consists of spotty information

---

73. I said, in section 4, how to turn a theory into an implicit definition of its basic quantities: take the function $o(T_1...T_n ; Q_1...Q_n)$ defined by a theory, where $T_1...T_n$ are theoretical properties, and $Q_1...Q_n$ qualities; replace each of the theoretical properties by variables $\exists t_1...t_n \ o(t_1...t_n ; Q_1...Q_n)$, and then implicitly define each T_i as the unique entity satisfying the open formula obtained by removing the existential quantifiers binding t_i. The difference between implicit and explicit definitions is that in the latter case one replaces the defined term by a formula in the language of the theory. Whether an implicit definition can be turned into an explicit one depends on the logical framework in which one is working (only with a sufficiently strong background language can it always be done), but the difference is one of some technical subtlety and isn't important here.

Science and Symmetry 183

about the actual distribution of observable quantities in one part of the universe. We embed this in models with higher-level structure, where the observable pattern is regarded as the causal product of the configuration of a small stock of underlying properties with a much nicer distribution (bearing one another the kinds of simple, local, invariant functional relations that can be discerned and projected). The underlying properties are not themselves observable, but they are picked out by rigidified causal description in qualitative (i.e., observable) terms. The descriptions are complicated; they are given by taking the theory as an implicit definition of the whole network of quantities and then picking out each quantity by its place in the network.[73] It is distinguished from more influential and familiar accounts of theorizing in its emphasis on the choice of basic quantities and the suggestion that most of the important theoretical decisions are made there. I wanted to add here only the observation that our theories not only have empirical content which transcends anything contained in the evidence for them, but a modal force which transcends even this content; in Cartwright's terms, the basic parameters of our theories are not just quantities but capacities.

7. PROPERTIES

(a) physical theorizing as decomposition of properties

The thought that physical theorizing involves analysis is not a new one, but it is most often accompanied by a conception of analysis that is tied too closely to the physical operation of breaking a large concrete thing into smaller parts; it is conceived, that is, as an analysis of *entities* into their spatially localized constituents. The notion of analysis more appropriate to physics is a mathematical one, and it applies primarily to the *properties* of things. I have been suggesting that the most important work in theorizing goes into the choice of the basic quantities, that it is made with an eye to teasing out hidden regularities between properties which can be translated back into qualitative regularities to derive predictions about the course of experience, and that the process is driven by a

---

74. 'Ascertainable' and 'characteristic' both in the senses of section 5.6.

combination of epistemic aspirations and limitations. We want to be able to predict the observable features of the evolution of systems on the basis of presently accessible knowledge, so what we do is cast around for a way of dividing systems into ascertainable types which evolve in a characteristic manner, so that we can decide the type of any system by empirical test and determine therefrom how it will evolve under arbitrary conditions.[74] In the terminology of sections 5.5 and 5.6, we cast around for a set of ascertainable properties whose distribution is both *characteristic* and *nice*.

If the most familiar sort of analysis is analysis of a spatially extended thing into spatial parts, a more abstract but still familiar sort, is analysis of a temporally extended thing into a collection of temporal parts. Combining the two, and taking them to the extreme, a spatially extended thing is analyzed into point-sized parts, and its history into a continuous series of instantaneous states. A less familiar sort of analysis decomposes the properties of a spatially extended system into those which are not intrinsic to any proper spatial part of it. This is the idea behind Fourier analysis of wave motion, for example, waves are decomposed into a set of 'modes', dynamical equations governing the evolution of the individual modes are given, and rules are provided for obtaining the motion of the whole from that of its parts.[75]

It is time now to make explicit what I have supposed about the nature of properties and bring together some of the things that I have said about them along the way. I have relied on an extensionalist conception of properties, made a distinction between qualitative and non-qualitative properties, and suggested that the basic quantities of a physical theory are implicitly defined in qualitative terms by the theory itself. I will take each of these up in turn.

(b) what properties are

The intrinsic properties of a physical system at a time are determined by a complete specification of the values of the physical quantities pertaining to the system at that time. Among these, the relatively abiding ones (e.g., mass, charge, spin-type)

---

75. Even if the wave is governed by non-linear equations, we may be able to approximate it by regarding the evolution of the individual modes as interdependent, though the interaction between them can very quickly can become so complicated as to be intractable.

determine its type and the rest go into the specification of its state, and our physical theories tell us how these properties relate to one another, how they evolve over time, and how they interact with the properties of other systems.

The history of philosophical discussion of properties is long and distinguished; it goes from Plato, Locke, Berkeley, and Hume, through Russell, Quine, and Goodman, to Putnam, Armstrong and David Lewis. Like all such discussions, it has been fraught with controversy and the more of that which I can sidestep, the better. Fortunately, much of it can sidestepped by focusing on the relatively precise notion of property involved in talk of the states and types of systems as they figure in scientific contexts and ignoring any extra-scientific concerns one might want an account of property to answer to. As I have suggested it should be understood, the notion of property needed to make sense of such talk is *extensionalist*. Physical properties are arbitrary sets of physical individuals; they have no essences, any way of picking one out is as good as any other. Combined with agnosticism about the existence of universals, tropes, natural kinds and other metaphysical entities to which sets may bear a distinguished relation and which philosophers have variously proposed as necessary conditions for genuine propertyhood, extensionalism is the most metaphysically innocent of the many philosophical accounts of properties that have been offered.

---

76. The set of materialistically possible individuals, individuals at worlds obtained by recombination of properties instantiated at the actual world, are different from the set of metaphysically possible individuals, as the latter are usually understood because of the metaphysical possibility of 'alien' properties. These are either simple properties which are instantiated nowhere in the actual world, or complex properties having as constituents simple properties instantiated nowhere in the actual world. Such properties are, by their nature, hard to find examples of, but we can pretend 'being a unicorn' is one.

If there are alien properties, then the materialistically possible worlds are only a proper subset of the metaphysically possible ones. Moreover, alien properties cannot be individuated by their extensions at materialistically possible worlds; simple alien properties have the same extensions there, viz. the null set. Mixed properties consisting of the conjunction or disjunction of any alien property and some physical property (e.g., (x(x is water = x is an aggregate of H_2O molecules v x is a unicorn)) are extensionally equivalent to the physical property by itself at such worlds, and so if we are concerned with individuating alien properties, we will have widen our gaze beyond the set of materialistically possible ones. In physical contexts, however, we can restrict our attention to the materialistically possible worlds, since we are only interested in non-alien properties and they can be individuated by their extensions within this smaller circle.

In saying that physical properties are just sets of physical individuals, I didn't mean actual individuals, for then it would make no sense to speak of distinct properties such that all actual things either have both or neither. Nor did I mean actual and physically possible individuals, for then there would be no such thing as distinct properties which are conjoined as a matter of law. I meant, rather, that properties were sets of materialistically possible individuals: P_1 and P_2 are the same property *iff* for all x ($P_1(x)=P_2(x)$), where x ranges over individuals in all such worlds.[76] When we ask whether a given temperature is the same physical property as some value of mean kinetic energy, we are asking whether the two vary independently of one another in the full circle of materialistically possible worlds.[77]

The empirical content of this modal talk lies partly in the consequences it has for which properties vary independently of one another in the actual world, but that isn't the whole story, for distinct properties which covary as a matter of law never actually occur apart. The empirical content of the latter distinction is a difficult to pinpoint and a little off the main line of argument, but contrary to a common charge[78], an extensionalist account does have the semantic resources to make it out.

The extensions of distinct quantities which covary with one another as a matter of law have the same extension in all *physically possible worlds*, but come apart in the wider circle of materialistically possible worlds. The distinction between reducing temperature to mean kinetic energy and regarding the one as a function of the other is that only in the latter case will there be materialistically possible worlds where the two have different values.

So much for extensionalism; I'll turn now to the distinction between qualitative and non-qualitative properties. Among all of

---

77. It is worth pointing out that the fact all and only sets of possible individuals are physical properties, and hence that all and only partitions of possible individuals are physical quantities, does not make the problem of determining what physical quantities there are trivial, because it is the choice of basic quantities which determines the set of possible individuals by delimiting *via* recombination the set of materialistically possible worlds.

78. To which Putnam, here, gives voice:
"if we accept the strict extensionalism which is urged by Quine, then all questions of reduction of properties trivializes upon the passing-over to corresponding questions about sets....[e.g. take the case of whether temperature is mean kinetic energy] one can take the relation extensionally as meaning that temperature is a one-one function of the equivalence-classes, subject to a continuity condition; but then *one will not have distinguished between the cases in which one magnitude is a function of another, and the cases in which one magnitude reduces to another*, which is just our problem." ("On Properties", p. 315)

the properties there are, there is a special subset which are distinguished by their epistemological relation to us; the properties to which our senses are attuned, whose distribution makes up the phenomenal structure of the world. These are the qualities. I introduced the distinction between qualitative and non-qualitative properties in the discussion of observable/unobservable distinction and invoked it again in connection with the Russellian response to Goodman's puzzle about induction. I'll elaborate here on what I said there and add some qualifications.

(c) qualitative and non-qualitative properties

Our senses are attuned to qualities in the sense that two objects look different *iff* they are qualitatively different. These properties are the qualities and two objects look different (under canonical conditions to ordinary observers) *iff* they are qualitatively different. The relations of resemblance in appearance—and consequently the distinction between qualitative and non-qualitative properties—are given directly in experience (indeed, I would say that they are *all* that is given in experience), and a part of the epistemic basis to which an empiricist may freely appeal. What distinguishes qualities from non-qualitative properties? Is it simply a brute difference in how things look to us, or is it grounded in some ontologically more basic distinction? I needn't say. Certainly there is no more *epistemically* basic fact than that the world presents itself to the senses carved into natural groupings; vision into kinds of colors, hearing into sorts of sounds, and so on. It is indisputable, moreover, that we all more or less agree on the groupings. In the most striking cases, the judgment that one collection of things appears kindred and another motley, is immediate and irresistible. If one is taken into a room filled with a large number of blue and green balls, and told to sort them into kinds, there is no question, if she is one of us, in the Wittgensteinian sense, that she will put the blue balls in one pile and the green in another.[79] If she is introduced to the predicate 'green' by exhibiting three or four green balls and contrasting them with some of blue ones, she will know instantly to which of the remaining balls in the room 'green' applies.

The thing is so natural that one has to work oneself into a strange frame of mind to be puzzled by it, but puzzling it is, if one regards any collection of balls as a property and any way of sorting

---

79. Assuming that they are all the same size, shape, and so on.

them as a partition into classes which share a property, for if we are sorting n balls into m piles, there are m•n different assortments, and hence m•n different families of properties into which they can be sorted. Each time we place a ball in either pile, we narrow the set of remaining assortments, but there is never a point in the process at which we have narrowed it enough to decide the pile into which the *next* ball should fall ... and yet we need scarcely to be told to sort balls by kind to know that it is the blue/green assortment that is wanted. Of all the possible ways of sorting the balls, only the one which places the blue ones together and the green ones together places each with those that resemble it in appearance, with those that are, that is to say, qualitatively similar to it. Puzzling or not, and grounded or not in some intrinsic difference between qualitative and non-qualitative properties, the distinction is real, and I haven't been shy about appealing to it ((i) in distinguishing observable structure described by our physical theories from the unobservable structure, and (ii) in distinguishing properties which can be ostensively identified [identified simply by displaying positive and negative instances] from those that cannot). It is important, though, not to assume too crude an understanding of it; I don't want to assume, for instance, that the natural groupings are perfectly well-defined and fixed, invariant across people and stable over time. There is a temptation to reason as follows: since the qualities are supposed to provide a minimal basis for characterizing experience, perceptual fields obtained from one another by exchanges among qualitatively similar regions should be indistinguishable. This has suggested to some that we might be able to use perceivers' judgments about the most fine-grained discriminations they can make between perceptual fields to identify the qualities; visual qualities, for example, would be sets of interchangeable parts of the visual field, auditory qualities would be sets of interchangeable sounds, and so on. Carnap, in the *Aufbau*, and Goodman, in *The Structure of Appearance*, set themselves (something like) the project of picking out the qualities in this way, but even before the daunting formal difficulties that faced those projects, there is a problem with the very idea.[80] Namely, that as soon as one goes beyond similarity in the most salient visual respects, judgments about the similarity of perceptual fields (or dispositions to

---

80. For a nice discussion of the formal difficulties, see Michael Dummett, *Review of Goodman's Structure of Appearance*, in *Truth and Other Enigmas*.

such judgments) get soft; it becomes clear that there is not really a well-defined stock of properties which is invariant across people and stable over time, and with respect to which experience can be described in full and final detail. Nor is there any reason to think that there are conditions (specifiable in a non-question-begging way) which, if they obtained, would put us each in a position to grasp experience in full and final detail, and that our experience, under these conditions would be the same.

Cases like the following point in the opposite direction. Suppose that we introduce Brenda to a set of wines telling her that they are similar, they all share the property of being "peesh", and suppose that at first they seem to her like a motley collection; she is unable, that is, to discern wherein their similarity resides, and has no idea how to sort new wines into or out of the collection. After some experience with wines and schooling with experts, however, her training 'takes'; she begins to sense their similarity and she is able to tell when previously untasted wines belong with them. She now knows, that is to say, what peeshiness is, and her finding out wasn't just a matter of learning that peeshiness was some construction of qualities with which she was already acquainted. Whereas at the outset of her training, to tell Brenda that a wine was peesh didn't raise expectations in her about how it would taste, the information now might be rich and important. She might now pay $150 for a wine on the strength of its peeshiness; she might now know to combine it with sole, and refrain from serving it with sweet deserts; she might now know just what kind of cheese it should accompany.

Connoisseurship in taste, smell, listening... are (at least in part) about 'refining' one's senses in this way, teasing out particular qualities and learning to appreciate, in the words of M.F.K. Fisher, 'differentiations in flavor' to which one was previously insensitive. The thing to notice is that, aside from the difference that training can make to one's own qualitative apprehension, there is no guarantee that, even with training, all of us can made be made to make equally fine discriminations, or even, for that matter, the same ones. In fact, snobbery about these matters depends on there *being* no such guarantee. M.F.K. Fisher writes, fore example, concerning taste:

> Almost all people are born unconscious of the nuances of flavor. Many die so. Some of these unfortunates are physically deformed, and remain all their lives as truly taste-blind as their brother sufferers are blind to color.

Others never taste because they are stupid, or, more often, because they have never been taught to search for differentiations of flavor.

They like hot coffee, a fried steak with plenty of salt and pepper and meat sauce upon it, a piece of apple pie and a chunk of cheese. They like the feeling of a full stomach. They resemble those myriad souls who say, "I don't know anything about music, but I love a good rousing military band."

Let the listener to Sousa hear much music. Let him talk to other music-listeners. Let him read about music-makers.

He will discover the strange note of the oboe, recognize the French horn's convolutions. Schubert will sing sweetly in his head, and Beethoven sweep through his heart. Then one day he will cry. "Bach! By God, I can hear him! I can hear!

That happens to the taste-blind in just some such way...It might be good if you could go to them, quietly, and say, "Please, sir, stop a minute and listen to me. Can you imagine eating bananas and Limburger cheese together? You have never thought about it? Then think. Taste them separately in your mind, the banana, the Limburger. Taste them together. Ah! It is horrible? Then how about mutton chops with shrimp sauce? And try herring soup with strawberry jam, or chocolate with red wine. (M.F.K. Fisher, *The Art of Eating*, p. 57)

What this suggests is that we have to recognize that our repertories of qualities are, to some extent, personalized (mostly they overlap; the salient visual properties that play the most important role in scientific contexts, in particular, are almost wholly shared), and that being able to discern a given quality, i.e. to apprehend what some set of objects 'have in common' and project it so that we know *what it would be like* for an as yet unexamined object to share it, is a skill. It requires training and attention, and accepting a theory in which they play an important role may be instrumental in focusing attention and motivating training.[81] None of this

---

81. Suppose I develop a 'theory' of wine that represents each as an admixture of five 'pure tastes' (one of which is peesh), what one obtains by mixing five 'eigen-wines' in different proportions, and that people can be brought by the right kind of training to taste the kindredness of peesh wines, to see what distinguishes them, as a group, from wines with no peeshy admixture, and even to distinguish by taste mixtures that previously seemed to them for all the world like qualitative duplicates. It might be only as a result of the development of the theory that discriminations in respect of peeshiness ever get made, but once apprehended, they become a part of the qualitative data which provides the evidential basis for any theory, and which any theory must accomodate, on pain of empirical inadequacy.

threatens the idea that there is a distinction between qualitative and non-qualitative properties, between properties which can be discriminated by the senses and those which cannot, for we can surely hold that there is such a distinction though it be not constant or shared, or anywhere etched in stone.

There is one worry about the use I've made of the distinction that needs to be addressed in light of its softness. I said in section 6, that the basic quantities of a theory provide an *adequate* partition only if *all* qualitative differences are represented by differences in the theoretical properties to which they are reduced; only in this case does a complete qualitative description of the world fall out of its theoretical description. But if what I just pointed out is correct, i.e. if there is no fixed stock of qualities invariant across people and over time, there is no single complete list of qualities against which the adequacy of a theoretical partition can be measured. It is clear that differences in color and temperature must be accommodated, but what of differences in the humorous properties of things (e.g. whether some string of symbols on a page is funny), or differences in the smells of cigars detected only by a rare breed of *aficionado*? Is our physics is responsible for representing these distinctions?

I am inclined to think that the situation must be something like what Putnam is suggesting in the following passage:

> each of the scientific disciplines has pretty much its own list of 'fundamental magnitudes' ... In each case there seem to be certain magnitudes which are 'dependent', in the sense that it is the business of the discipline (a discipline may, of course, change its mind about what its 'business' is) to predict their time-course, and certain magnitudes which are independent, in the sense that they are introduced in order to help predict the values of the dependent magnitudes. In physics, for example, it was the position of particles that was, above all, dependent. In economics it would be prices and amounts of production. In each case the scientist looks for a set of properties including his 'dependent variables' ... which will predict, at least statistically, the time-course of the dependent variables... (Putnam, "On Properties", p. 319)

In developing a science, we start out with a set of properties we'd like to be able to predict and control; these turn out to be causally connected to other properties in such a way that prediction and control of the former require attention to the latter. The humorous properties of things don't start out on our list as dependent variables in any science, nor do they show up there later in virtue of being causally integrated with the things that *do* start out on our list; they are not, that is, introduced in the course of theorizing about those as independent variables. Most of us complacently suppose that humorous properties will turn out to supervene in some complicated way on whatever our basic physics takes as the fundamental quantities, for we are impressed by the very pretty fact that the dependent variables of all other sciences seem to do so. But it wouldn't be a disaster for our physics if they didn't turn out to (in the way it *would* be a disaster for our physics if we found that colors didn't supervene); it wouldn't, that is, overturn our very ideas about what the basic physical quantities were.[82] If the causal closure of the salient qualitative differences it is the business of physics to predict doesn't include the humorous properties of things, and the humorous properties of things turn out not to supervene on those that it does, then perhaps we would regard the humorous properties of things as 'non-physical'. Mapping the structure of the world with respect to its humorous properties would still retain whatever intrinsic interest it has, but it would not be an urgent scientific project, and given the lack of causal interaction with physical variables, I don't suppose the scientific world would be shaken by the discovery of the existence of non-physical properties.[83] What this suggests is not that there is no set of qualities to which the basic theoretical partition is required to be adequate, but that physics itself may play a partly legislative role in determining its composition, and this is quite compatible with holding that there is some set of qualitative distinctions against which the adequacy of our basic theoretical partition is

---

82. That, at least, is the conceit of physics. If it is warranted, to deny that humorous properties supervene on the independent variables of physics (particle positions, spin values, and whatnot) is to deny that they are causally integrated with any of these other variables; they are not, that is, included in the causal closure of the dependent variables of the sciences.

83. A similar story could be told about moral or aesthetic properties. Either we suppose they supervene on the independent variables of physics without having any clear idea about how (i.e. without having any idea about how to reduce the one to the other), or we deny that they are physical. It used to be thought that we knew what the basic physical quantities had to be, and so we had an easier time giving content to the denial that some set of properties were physical; it was just to deny that they supervened on those. For Descartes, for example, to deny that a property was a physical one was to deny that it supervened on the values of the kinematical quantities. The idea that we can determine *a priori* the basic quantities, however, was tossed out the window early this century, and with it the simple criterion for the 'physicality' of properties.

measured. It relaxes, and deepens our understanding of, the constraint, but does not, I think, force us to relinquish it.

(d) ascertainable and unascertainable quantities

The values of observable quantities are qualitatively different, they can be distinguished by unimplemented observation; the values of unobservable, measurable quantities can be distinguished by observation implemented with an imaging instrument or a measuring apparatus. I call both 'ascertainable' because whether they obtain can be determined empirically; there are conditions (themselves ascertainable) under which observation—unimplemented in the one case, implemented in the other—functions as a test for their presence. Not so for unascertainable quantities; no amount of observation, however implemented, will serve to determine their distribution. The ascertainable properties provide the sparse notion of property which (David Lewis has influentially argued) is essential to accounts of interpretation, causation, lawhood, counterfactuals, content, and so on.[84] Like the distinction between

---

84. Lewis' arguments are given in their most complete form in "New Work for a Theory of Universals", *Australasian Journal of Philosophy*, vol. 61, no.4, Dec. 1983 (but see also "Putnam's Paradox"). He writes in the former:

> "Because properties are so abundant, they are undiscriminating. Any two things share infinitely many properties, and fail to share infinitely many others. That is so whether the two things are perfect duplicates or utterly dissimilar. Thus properties do nothing to capture facts of resemblance. ... Properties carve reality at the joints - and everywhere else as well. If it's distinctions we want, too much structure is no better than none. It would be otherwise if we had not only the countless throng of all properties, but also an elite minority of special properties... [these] would be the ones whose sharing makes for resemblance, and the ones relevant to causal powers."

> Lewis is undogmatic about wherein lies the specialness of the special properties; he opts for a primitive property of 'naturalness', but acknowledges that they could be picked out by universals or that taking relations of resemblance among individuals as primitive, and then defining natural classes as those whose members resemble one another in a particular respect would also work.

> I am following a slightly different strategy by appealing to relations of resemblance *in appearance*, i.e. qualitative similarity, to give a substantive account of science, and then allowing genuine physical similarity to consist in the sharing of properties which are the values of the quantities of our physical theories. These are required to be ascertainable and it is these which are relevant to a system's causal powers.

qualitative and non-qualitative properties on which it relies, the distinction between ascertainable and unascertainable quantities is epistemic. It is just the distinction between properties we can detect (given unlimited time and technological resources, and perfect knowledge of the laws) and those we can't, and should be acceptable to Platonists and nominalistically minded philosophers, alike. These are the only distinctions among properties that are relevant to (and hence play any role in an account of) physics. Perhaps we can infer *from* the sharing of values of physical quantities *to* metaphysical distinctions (e.g., natural kindhood, correspondence to a universal) but these aren't distinctions we rely on in picking out the physical quantities.

If we take a syntactic formulation of a theory, replace all predicates which denote non-qualitative properties with variables $*(a_1, \ldots a_n)$, we get an open sentence $F(a_1, \ldots a_n)$, where the only non-logical vocabulary in F are predicates denoting qualities. Binding it with the appropriate quantifier, we obtain a description of a network of non-qualitative properties in qualitative terms: "the $(a_1, \ldots a_n)$ such that $F(a_1, \ldots a_n)$", i.e. 'the network of properties which are actually related in such and such a way to qualities'. The network is picked out by its causal relations to qualities, and individual properties are picked out by their place in the network.[85] It is a consequence of the view discussed in section 5.4 about how the higher-level structure of the models of our theories is interpreted, that our physical theories function to implicitly define their basic quantities in this way. If what I have said is right, our physical theories have a dual role: they simultaneously provide implicit qualitative definitions of their basic quantities, and specify the range of structures which represent their physically possible distributions.

(e) qualities in the world of physics and the phenomenal world

---

[85]. There are two qualifications: the description is guaranteed to be satisfied by *some* set of properties provided only provided that it doesn't mistake the cardinality of the world, and it is guaranteed to be *uniquely* satisfied only so long as the network doesn't possess any troublesome symmetries, i.e. provided that it is not symmetric with respect to exchanges among any set of properties.

[86]. 'We' here, refers to everyone who shares my perceptual apparatus and my quality space.

From a physical point of view, what sets qualities apart from other properties is that we bear them a special relation (namely, we can tell whether they obtain just by looking); from a physical point of view the distinction between qualitative and non-qualitative properties is thoroughly anthropocentric.[86] Because of the way in which physical properties are causally integrated with one another, we typically have an indefinite number of very different tests for the presence of any one. In general, if the interaction between two systems X and Y has the effect of correlating an observable Y-quantity with a particular quantity on X, letting X and Y interact and watching for the value of the relevant Y-quantity counts as a test for the presence of any value of the X-quantity in question; it counts, that is, as a measurement of the X-quantity. What makes greenness a quality is that we can tell whether an object is green just by looking at it, but there is nothing special about looking; it is, like any other measurement, just test for the presence of greenness.[87] Greenness doesn't have an essence that is wholly and completely grasped in looking at green things (and only in looking at green things); we don't come to know what green is—as David Lewis recently put it—in an 'uncommonly demanding and literal sense of knowing what' by having a green experience.[88] From a physical point of view, we don't learn any more about greenness by seeing green things than we do by passing the light reflected by them through a spectrometer.

From an epistemological perspective, however, qualities have a very special status; if what I have said about them is right, they are also the touchstone of physical understanding and informativeness. The proposition that an object possesses some quality is informative in itself (e.g., for the trained taste of Brenda, the proposition that a wine is peesh carries information about the wine); the proposition that an object possesses some non-qualitative property, on the other hand, is not; it tells us nothing until we have a theory in which its causal connections with qualities are specified, so

---

87. This might be because (with all the relevant qualifications and idealizations) looking-green-to-us (or, better, looking-to-us-like-other-green-things-we-have-seen) is just what it is for something to be green.

88. Lewis isn't endorsing this view; it is a part of the folk-psychological notion of greenness that he articulates to reject (wearing the hat of a materialist). The article is "Should a Materialist Believe in Qualia?", *Australasian Journal of Philosophy*, vol. 73, no. 1, March 1995.

89. I have been assuming, though I haven't been careful to say so, n-place spatial and temporal relations are qualitative properties of n-tuples.

that we can draw conclusions about causally associated qualitative properties therefrom. Meaningful discourse about the physical world bottoms out in its qualitative content.[89]

In the world-picture embodied in our physical theories, the phenomenal world appears as a coarse-grained description of a world which is characterized at a more basic level by the fundamental physical quantities and relations, and the qualities whose distribution characterizes the phenomenal world appear as logically complex constructions out of these. In the phenomenal world-picture, by contrast, the physical quantities and relations themselves appear as abstractions from the simple qualities which are the atoms of experience. Our physical theories paint a picture of the world (replete with us as a part of it, picture-painting as we are wont to do, and even theorizing about our picture-painting), *from the outside, in*, in terms of the properties that our theories take as basic and from which the materials we employ in our own picture painting (as depicted) must be constructed. Our epistemological theories, on the other hand, paint a picture of this process of picture-painting (in all of its detail, including the part which depicts our picture-painting and also our theorizing about our picture-painting), *from the inside, out*, in terms of the properties which are the simples of experience and from which we 'factor out' the basic quantities of our theories.

Each of the pictures is internally consistent, but they differ in the relative priority of their constituents, over which quantities are taken as basic and which derived. More than one philosopher has seen in them opposing visions and thought of them as presenting a dilemma; if we can't find a way of fitting them into a common frame (and there has been little recognized success in doing *that*), we must reject one in favor of the other. The assumption behind the dilemma seems to be that insofar as it is a single world we are trying to depict, there has to be a single complete and correct picture of it. Seeing what is wrong with the assumption would lead us into a discussion of some larger issues, for it would lead us into a general discussion external relations between different world-pictures, and if I were to enter it, I would take a leaf from Goodman's book and suggest that we should widen our gaze beyond the narrow scope on which we have kept it focused so far to include not only the world-pictures embodied in the various sciences and in experience, or even the vague and various pictures embedded in common sense, but the often spotty and partial visions embodied in works of art; linguistic and non-linguistic, representational and abstract. I would agree with Goodman that the world-pictures of physics and experience, like these others, don't compete, but coexist, even complement, perhaps even require, one another.

Such a discussion would also be useful because it would dispel the impression I may have left with all the talk of the kind of comprehensiveness which physics claims for itself, that I am falling in line with physical fundamentalists, i.e. that I am endorsing the view that physics provides *the* complete picture of the world, that *all* (and only) the real world's inhabitants appear therein, and so—to put it a little paradoxically—nothing that doesn't appear therein is real.[90] On the contrary, I think that physics provides one vision of the world among others. It is one with an unparalleled combination of precision, beauty, and scope, but if I have spoken as though it provides the full and final version in which all others must be embedded or rejected, that is only because that is a condition of adequacy that physics imposes on itself, rather than one that derives from an external ordering of world-versions which puts physics at the base We can, in elaborating the world of physics, that is to say, describe, without endorsing, its claim to hegemony. The discussion of art, were it to be given, would also provide the background against which the kind of activity involved in scientific theorizing, as I've represented it, could be seen as a special case of a more general one—'rendering'—that also characterizes the arts. But it is, for the time being, merely hypothetical.

---

90. Of course, nothing that doesn't appear in that picture is *physical*, but that is just to say that nothing that does not fit into the world picture of physics fits into the world picture of physics. I'm not endorsing the identification of the physical with the real.

Bibliography

Adler, R. Bazin, M., and Schiffer, M., *Introduction to General Relativity*, New York: McGraw-Hill (1965).

Anderson, J. L., *Principles of Relativity Physics*. New York: Academic Press (1967).

Armstrong, D., *What is a Law of Nature?*, New York: Cambridge University Press (1983).

Arnold, V. I. and Avez, A., *Ergodic Problems of Classical Mechanics*, New York: Benjamin (1968).

Bishop, R. I., and Goldberg, S. I., *Tensor Aalysis on Manifolds*, New York: Macmillan (1968).

Black, M., "Models", in *Models and Metaphors*, Ithaca, N.Y.: Cornell University Press (1962).

———. "The Direction of Time", in *Models and Metaphors*, Ithaca, N.Y.: Cornell University Press (1962).

Bohm, D., and Hiley, B., *The Undivided Universe: an Ontological Interpretation of Quantum Theory*, London: Routledge (1993).

Bohm, D., *Wholeness and the Implicate Order*, London: Routledge, (1980).

———. *Causality and Chance in Modern Physics*, London: Routledge & Kegan Paul (1957).

———. *Quantum Theory*, New York: Prentice Hall, (1951).

Boltzmann, L., *Lectures on Gas Theory*, Brush, S. (trans.), Berkeley: University of California Press (1964).

Boyd, R., "The Current Status of Scientific Realism", in J. Leplin (ed.), *Scientific Realism* (1984), 41-82.

Boyle, R., *The Works of the Honourable Robert Boyle*, ed. T. Birch, London (1673).

Carnap, R., *The Logical Structure of the World*, George, R. (trans.), Berkeley: University of California Press (1967).
———. *Philosophical Foundations of Physics*, ed. Martin Gardner, New York: Basic Books (1966).
Cartwright, N., *Nature's Capacities and Their Measurement*, Oxford: Oxford University Press (1989).
———. *How the Laws of Physics Lie*, Oxford: Oxford University Press (1983).
Chalmers, A.F., (1970) "Curie's Principle", *British Journal for the Philosophy of Science* 21, 133-148.
Churchland, P., and Hooker, C. (eds.), *Images of Science: Essays on Realism and Empiricism, with a Reply by Bas C. van Fraassen*, Chicago: University of Chicago Press (1985).
———. (1985) "Reduction, Qualia, and the Direct Introspection of Brain States", *Journal of Philosophy* 82, 8-28.
Curie, Pierre, (1894) "On Symmetry in Physical Phenomena, Symmetry of an Electric Field and of a Magnetic Field", *Journal de Physique* 3, 401.

Dalla Chiara, M. L., and Toraldo di Francia, G., "A Formal Analysis of Physical Theories, in G. Toraldo di Francia (ed.), *Problems in the Foundations of Physics*.
Davies, P.C.W., *The Physics of Time Asymmetry*, Berkeley: University of California Press (1974).
Dummett, M., review of "*The Structure of Appearance*", in *Truth and Other Dogmas*, Oxford: Clarendon Press (1978).

Earman, J., *World Enough and Space-Time: Absolute vs. Relational Theories of Space and Time*, Cambridge, Mass.: Bradford (1989).
———. "Laws of Nature: The Empiricist Challenge", in R. J. Bogdan (ed.), *D. H. Armstrong*, Dordrecht: Reidel (1986).
———. (ed.), *Testing Scientific Theories, Minnesota Studies in the Philosophy of Science*, x, Minneapolis: University of Minnesota Press (1984).
———. (1978), "The Universality of Laws", *Philosophy of Science*, 45, 173-81.
———. and Glymour, C., (1978) "Lost in the Tensors: Einstein's Struggles with Covariance Principles, 1912-1916", *Studies in History and Philosophy of Science* 9, 251-78.
———. (1974) "An Attempt to Add a Little Direction to the Problem of the Direction of Time", *Philosophy of Science* 41, 223-37.

———. (1974), "Covariance, Invariance and the Equivalence of Frames", *Foundations of Physics* 4, 267-89.
———. (1970), "Space-time, or How to Solve Philosophical Problems and Dissolve Philosophical Muddles Without Really Trying", *Journal of Philosophy* 67: 259-77.
———. (1971), "Kant, Incongruous Counterparts and the Nature of Space and Space-Time", *Ratio* 13, 1-18.
Ehrenfest, P. and T., *The Conceptual Foundation of the Statistical Approach to Mechanics*, Moravscik, M. (trans.), Ithaca, N.Y.: Cornell University Press (1959).
Einstein, A., *The Meaning of Relativity*, 5th ed. Princeton, N.J.: Princeton University Press (1953).
———. "The Foundation of the General Theory of Relativity", in Perrett, W., and Jeffery, G. (trans.), The Principle of Relativity: A Collection of Original Memoirs on the Special and General Theory of Relativity, New York: Dover (1923).

Fermi, A., *Thermodynamics*, New York: Dover (1936).
Feyerabend, P., "An Attempt at a Realistic Interpretation of Experience", in *Philosophical Papers*, 2 vols., Cambridge: Cambridge University Press, (1981).
Feynman, R., *The Character of Physical Law*, Cambridge: MIT Press (1965).
Field, H., *Science Without Numbers*, Princeton, N.J.: Princeton University Press (1980).
Friedman, M., *Foundations of Space-Time Theories: Relativistic Physics and Philosophy of Science*, Princeton, N.J.: Princeton University Press (1983).
———. (1982), "Review of van Fraassen, *The Scientific Image*", *J. Phil.* 79, 274-83.
———. (1979), "Truth and Confirmation", *J.Phil.* 76, 361-82.
———. (1974), "Explanation and Scientific Understanding", Journal of Philosophy 71, 5-19.
———. "Relativity Principles, Absolute Objects, and Symmetry Groups" in Suppes, P. (ed.) *Space, Time, and Geometry*, Dordrecth: Reidel (1973).

Geroch, R., *Mathematical Physics*, Oxford: Oxford University Press (1980).
Giere, R., *Explaining Science*, Chicago: University of Chicago Press (1988).
———. "Constructive Realism", in P.M. Churchland and C. Hooker (eds.), *Images of Science* (1985).

———. *Understanding Scientific Reasoning*, New York: Holt, Rinehart, and Winston, (1979).
Glymour, C., "Explanation and Realism", in P.M. Churchland and C. Hooker (eds.), *Images of Science* (1985).
———. *Theory and Evidence*, Princeton, N.J.: Princeton University Press (1980).
———. "Topology, Cosmology and Convention", in Suppes, P. (ed.) *Space, Time, and Geometry*, Dordrecth: Reidel (1973).
———. (1972) "Physics by Convention", *Philosophy of Science* 39: 322-40.
Goodman, N., *Fact, Fiction, and Forecast*, Cambridge: Harvard University Press (1979).
———. *Ways of Worldmaking*, Indianapolis: Hackett (1978).
———. *Languages of Art*, Indianapolis: Hackett (1977).
———. *The Structure of Appearance*, New York: D. Reidel (1951).

Hacking, I., "Do We See through a Microscope?", in P.M. Churchland and C. Hooker (eds.), *Images of Science* (1985).
———. *Representing and Intervening*, Cambridge: Cambridge University Press (1983).
Harman, G., (1965) "The Inference to the Best Explanation", *Phil. Review* 74, 88-95.
Hempel, C., *Aspects of Scientific Explanation*, New York: Free Press (1965).
Horwich, P., *Asymmetries in Time*, Cambridge, Mass.: Bradford (1987).

Leplin, J. (ed.), *Scientific Realism* Berkeley: University of California Press (1984).
Lewis, D., *Philosophical Papers*, 2 vols., Oxford: Oxford University Press (1986).
———. (1983), "New Work for a Theory of Universals", *Australasian J. Phil.* 61, 343-77.
———. Mach, E., *The Science of Mechanics*, T. McCormack (trans.), LaSalle, Ill.: Open Court (1942).

Maxwell, G., "The Ontological Status of Theoretical Entities", in Feigl, H., and Maxwell, G. (eds.), *Minnesota Studies in the Philosophy of Science*, vol. III, Minneapolis: University of Minnesota Press (1962).

Nerlich, G., "Hands, Knees, and Absolute Space", in *The Shape of Space*, Cambridge: Cambridge University Press (1976).

Bibliography

Putnam, H., "Models and Reality", in *Realism and Reason*, New York: Cambridge University Press (1983).

———. "Review of Nelson Goodman's Ways of Worldmaking", in Realism and Reason, New York: Cambridge University Press (1983).

Quine, W., (1970) "On the Reasons for Indeterminacy of Translation", *Journal of Philosophy* 67, 178-83.

———. "Natural Kinds", in Ontological Relativity and Other Essays, New York: Columbia University Press (1969).

Reichenbach, H., *The Direction of Time*, Reichenbach, M., and Freund, J. (trans.), New York: Dover (1957). (original German version 1928).

———. *The Philosophy of Space and Time*, Reichenbach, M. (trans.), Berkeley: University of California Press (1956).

———. *Experience and Prediction*, Chicago: University of Chicago Press (1938).

Remnant, P., (1973) "Incongruent Counterparts and Absolute Space", *Mind, New Series* 72, 337-51.

Russell, B., *Human Knowledge: It's Scope and Limits*, New York: Clarion (1948).

———. "On the Notion of Cause with Applications to the Free Will Problem', in Feigl,, H. and M. Broadbeck, M. (eds.), *Readings in the Philosophy of Science*, New York: Appleton-Century-Crofts (1953).

Schrodinger, E., *Statistical Thermodynamics*, New York: Dover (1946).

Seager, W. (1996) "Ground Truth and Virtual Reality: Hacking vs. van Fraassen", *Philosophy of Science*.

Sklar, L., "Incongruous Counterparts, Intrinsic Features, and the Substantivality of Space", in *Philosophy and Space-time Physics*, Berkeley: University of California Press (1985).

———. *Space, Time, and Spacetime*, Berkeley: University of California Press (1974).

Skyrms, B., (1985) "Maximum Entropy as a Special Case of Conditionalization", *Synthese* 636, 55-74.

Suppe, F., *The Structure of Scientific Theories*, Urbana: University of Illinois Press, (1974).

Suppes, P., "The Structure of Theories and the Analysis of Data", in F. Suppe (ed.), *The Structure of Scientific Theories*, Urbana: University of Ilinois Press (1974) 266-83.

Tuller, A., *A Modern Introduction to Geometries*, Princeton, N.J.: van Nostrand (1967).

van Fraassen, Bas C., *Quantum Mechanics: An Empiricist View*, Oxford: Oxford University Press (1991).
———. *Laws and Symmetry*, Oxford: Oxford University Press (1989).
———. *The Scientific Image*, Oxford: Oxford University Press (1980).
———. *An Introduction to the Philosophy of Time and Space*, New York: Columbia University Press (1985).
———. "Empiricism in the Philosophy of Science", in P.M. Churchland and C. Hooker (eds.), *Images of Science* (1985).
———. "Glymour on Evidence and Explanation", in J. Earman (ed.), *Testing Scientific Theories*.
———. "Theory Comparison and Relevant Evidence", in J. Earman (ed.), *Testing Scientific Theories*.

Watanabe, S. "Reversibilite contre irreversitibilite en physique quantique", in *Louis de Broglie, physicien et penseur*, Paris: Albin Michel (1953).
Weyl, H. *Symmetry*, Princeton, N.J.: Princeton University Press (1952).
Wigner, E., *Symmetries and Reflections*, Princeton, N.J.: Princeton University Press (1965).

Author Index

Bennett, Johnathan, 64
Bohm, David, 175
Boltzmann, David 100

Carnap, Rudolph, 110, 188
Chalmers, Alan, 40
Churchland, Paul, 124
Curie, Pierre, 9

Demopolous, William, 146
Dummett, Michael, 188
Dyson, Freeman, 82

Earman, John, 64
Einstein, Albert, 80

Feyerabend, Paul, 126
Feynman, Richard, 82, 118
Fisher, M.F.K., 189–90
van Fraassen, Bas, 31–2, 147
Friedman, Michael, 7, 111, 146

Galilei, Galileo, 81
Gardner, Martin, 62, 76
Glymour, Clark, 6
Goodman, Nelson, 126, 152, 153–4, 162, 188

Hume, David, 151

Jordan, Pascual, 93

Kant, Immanuel, 61–84

Lewis, David, 193
Lie, Sophus, 55

Maxwell, Grover, 134

Nerlich, Graham, 64, 74

Pauli, Wolfgang, 78
Poincare, Henri, 48
Putnam, Hilary, 191

Radicati, Luigi, 33
Reichenbach, Hans, 152
Remnant, Peter, 64
Russell, Bertrand, 151, 152

Seagar, William, 135
Sklar, Larry, 64

Weyl, Hermann, 29, 82
Wigner, Eugene, 86
Wu, Madame, 83

Subject Index

Adequacy, **163–6**
*Ad-hoc*ness, 118
Ascertainability, 166–9

Boltzmann's Equation, 100, 104
Born's Rule, 44, 45
branch systems, 105

capacities, 180–1
cause, 33, 49, 54
change, 78
charity, in interpretation, **136–41**, 145–6
Collapse Postulate, 44
counterfactuals, 54, 179–83
covariance group, 6
Curie's Principle, 10, **31–58**

effect, 33, 49
empirical adequacy, 107, 109
enantiomorphs, **61–84**
Equiprobability Assumption, 100, 105
evidence, 136, 148
explication, 110

general covariance, 36
General Relativity, 79

group, 21

hidden variables, 47

imaging instruments, **127–34**
incongruous counterparts. *See* enantiomorphs
indeterministic contexts, 42, 91, 1319
individuals, 10
induction, **148–55**
    New Riddle of, 155
initial conditions, 86–90
interpretation, 109, **136–46**

laws, 3, 86–90
Lorentz contraction, 75

manifold, 8
mapping. *See* transformation
meaning, 77
measurement, 127
methodological rules, 14
microscopes. *See* imaging instruments
modality, 179–83
model:
    empirical equivalence of, 137
    in the logician's sense, 112

in the sense relevant to the semantic view, 111
motion, 79

Newtonian Mechanics, 79
niceness, as pertaining to theoretical structure, **170–75**

observation, 127
observational/theoretical distinction, **120–36**
ontology, 3

parity, violation of, 78
possibility:
    materialistic, 3, 185
    metaphysical, 185
    physical, 185
prediction, 179
properties:
    aesthetic, 191
    analysis of, 183
    ascertainable, 158, **192–4**
    characteristically distributed, 155, 184
    intrinsic, 59, 69, 184
    measurable, 128
    metrical, 63, 72
    moral, 192
    ostensively definable, 155
    positional, 156
    projectible, 155, 158
    qualitative, 126, 156, 158, **187–92**, 194–7

quantities:
    basic, 10, 129
    hidden, 12
    observable, 126
    unobservable, 133
Quantum Field Theory, 46

quantum mechanics, 44, 94

Ramsey sentences, **142–5**
randomness, 86, 89
reflection:
    in space, 15, 20, **57–84**, 90–93
    in time, 15, **94–106**
regularities:
    hidden, 171, 173
    manifest, 171
relations:
    external, 10, 6
    internal, 62
    local, invariant, 17

similarity, 57–8
simplicity, 119
space, 59
    Euclidean, 21
    mathematical, 78
    non-orientable, 67
    physical, 78
space-time:
    Minkowski, 75
    substantival, 15, 63
space-time theories, 5, **6–9**
spatiotemporal structure, 6
Special Relativity, 75, 79
spontaneous symmetry breaking, **45–7**
states:
    coarse, 43
    fine, 43
state-space, 19
structure:
    causal, 182
    observational, 10, 136
    qualitative, 16
    theoretical, 10, 136, 149
symmetries:

Index

apparent, 48, 53
characteristic, 34, 48
geometric, 19
hidden, 48, 53
idiosyncratic, 34
intrinsic, 51
of models, 9
of theories, 4
of worlds, 3, 9

Theoretician's Dilemma, 146
theories:
    coordinate-dependent formulations of, 6–9, 38, 41–2
    coordinate-independent formulations of, 41–2
    dynamical 55
    empirically equivalent, 107, 114, 146–7
    equivalent, 116
    observational content of, 113
    scientific, 107
    semantic conception of, 108
    syntactic conception of, 109
theory-ladenness, 121
thermodynamic states, 96
    macro-, 99
    micro-, 99
thermodynamics, **94–106**
topology, 68
transformation, 3, 19, 20
    geometric, 28, 55
    non-geometric, 28
    reversible, 97

Uniformity of Nature, 161

vocabulary:
    observational, 109, 144
    theoretical, 109, 144

weak nuclear interaction, 93